Nanomaterials for Advanced Energy and Power Storage Devices

Online at: https://doi.org/10.1088/978-0-7503-4901-7

Nanomaterials for Advanced Energy and Power Storage Devices

Edited by

K R V Subramanian

Department of Mechanical Engineering, Ramaiah Institute of Technology, Bangalore, India

N Sriraam

Department of Medical Electronics, Ramaiah Institute of Technology, Bangalore, India

T Nageswara Rao

Department of Mechanical Engineering, GITAM University, Bengaluru, India

Aravinda C L Rao

Reliance Industries Ltd, Vadodara, India

IOP Publishing, Bristol, UK

ISBN 978-0-7503-4901-7 (ebook)
ISBN 978-0-7503-4899-7 (print)
ISBN 978-0-7503-4902-4 (myPrint)
ISBN 978-0-7503-4900-0 (mobi)

DOI 10.1088/978-0-7503-4901-7

Version: 20241001

IOP ebooks

British Library Cataloguing-in-Publication Data: A catalogue record for this book is available from the British Library.

Published by IOP Publishing, wholly owned by The Institute of Physics, London

IOP Publishing, No.2 The Distillery, Glassfields, Avon Street, Bristol, BS2 0GR, UK

US Office: IOP Publishing, Inc., 190 North Independence Mall West, Suite 601, Philadelphia, PA 19106, USA

To our families.

Contents

Preface ix

Editor biographies xi

List of contributors xiii

1 Nanomaterials for static and dynamic flow supercapacitors 1-1
Soorya Sasi and Sunish K Sugunan

1.1 Introduction 1-1
1.2 Classification based on the working principle 1-3
1.2.1 Electric double-layer capacitors 1-3
1.2.2 Electrode materials of the EDLC 1-4
1.3 Pseudocapacitive electrode materials 1-6
1.3.1 Conducting polymers 1-6
1.3.2 Metal oxides 1-7
1.3.3 Nanocarbon-metal oxide composites 1-7
1.3.4 Hybrid supercapacitors 1-7
1.3.5 Classification based on the morphology 1-9
1.3.6 Static/rigid devices 1-10
1.4 Dynamic devices 1-10
1.4.1 Flexible supercapacitors 1-10
1.5 Materials used as substrates in flexible supercapacitors 1-12
1.6 Materials used for making the electrode 1-14
1.7 Materials used as the electrolyte 1-16
1.8 Electrochemical flow capacitor 1-18
1.9 Factors affecting the performance of the EFC 1-20
1.10 Conclusion 1-21
References 1-21

2 Emerging technological advancements in thermal energy storage 2-1
H S Balasubramanya, Niranjan Murthy and K R V Subramanian

2.1 Introduction 2-1
2.2 TES technologies 2-2
2.3 Performance and costs 2-6
2.4 Barriers to adoption 2-6
2.5 Applications of sensible thermal energy storage 2-6
References 2-10

3 Quantum dots as an efficient supercapacitor for the modern world **3-1**

Janardan Sannapaneni, H Manjunatha, K Venkata Ratnam and B Ramakrishna Rao

3.1 Introduction 3-1
3.2 Synthesis of carbon quantum dots 3-2
3.3 Synthesis of graphene quantum dots 3-3
3.3.1 Top-down approach 3-3
3.3.2 Bottom-up methods 3-3
3.3.3 Heteroatom doping 3-4
3.4 Carbon quantum dots for supercapacitors 3-4
3.5 Graphene quantum dots for supercapacitors 3-8
3.6 Conclusion 3-9
References 3-9

4 Importance of redox active electrolyte for the next generation energy storage system **4-1**

Gaytri, Partha Khanra, Pankaj Kumar and Sudesh Mittal

4.1 Introduction 4-1
4.1.1 Charge storage principles of redox electrolytes containing ECs 4-6
4.1.2 Mechanism of diffusion, adsorption, and kinetics of redox electrolytes 4-8
4.2 Categories of redox electrolyte 4-10
4.2.1 Aqueous systems 4-13
4.3 Non-aqueous redox electrolyte system 4-28
4.3.1 Ionic liquid-based redox electrolytes 4-28
4.4 Redox mediated gel-polymer electrolytes 4-31
4.5 Catholyte and anolyte-based redox electrolyte 4-32
4.6 Conclusion 4-34
References 4-35

5 Conclusion on current trends in energy storage systems **5-1**

Venkatesh Sadhana and Janardan Sannapaneni

5.1 Hydrogen energy storage system 5-5
5.1.1 Synthetic natural gas (SNG) 5-6
5.1.2 Solar fuels 5-6
5.1.3 Future aspects of energy storage systems 5-6
References 5-7

Preface

Electrical energy storage lies at the heart of human daily life, due to its wide range of applications in cell phones, laptops, cameras, iPads, and so forth. Rechargeable batteries are one of the favoured choices taking the place of the internal combustion engine by powering electric vehicles. In addition, there are enormous demands for grid energy storage to balance distribution and to solve the intermittent electricity supply generated by renewable energy resources, such as solar, wind, and waves.

Electrical energy storage devices are an integral part of telecommunication devices (cell phones, remote communication, walkie-talkies, etc), standby power systems, and electric hybrid vehicles in the form of storage components (batteries, supercapacitors, and fuel cells). In the field of energy storage, two main parameters are fundamental for storage devices: the energy density and the power density. The first parameter defines the amount of energy that can be stored in a given volume or weight while the second parameter describes the speed at which the energy is stored or discharged from the device. The ideal storage device should simultaneously have both high energy density and high power density. Hence, the integration of conventional primary energy storage units (e.g., batteries and fuel cells) and the electric energy storage devices in the high-power or pulse-power forms (e.g., capacitors) becomes the prime concern in the development of new power systems. On the other hand, the energy densities of conventional capacitors are usually too low to be acceptable for some future applications; the development of capacitors with high energy densities (i.e., supercapacitors) for these applications has become an exciting subject prompting much research into electrochemical energy storage/conversion systems.

Electrical energy storage (EES) systems are the key elements to the build-up of sustainable energy technologies. Electrochemical energy storage devices, which are known as supercapacitors or ultracapacitors, are categorized into various energy-storage processes, like electric double-layer capacitors (EDLCs) and pseudocapacitors. Firstly, electrical double layers are formed on the interface of the electrode and electrolyte, EDLCs store charge physically. This is an extremely reversible process, thus the cycle of life is really infinite. However, an electric double layer, fast surface oxidation–reduction reactions and possible ion intercalation in the electrode, store energy in energy storage systems (capacitor, pseudocapacitors). Supercapacitors are electrochemical energy storage devices that can be fully charged or discharged in seconds. Due to their higher power density, low maintenance cost, wide thermal operating range, and more extended cycle life compared to secondary batteries, supercapacitors have attracted significant research attention over the past decade. They also possess a higher energy density compared to conventional electrical dielectric capacitors. The storage capacity of a supercapacitor depends on the electrostatic separation between electrolyte ions and high surface area electrodes. However, the lower energy density of supercapacitors in comparison with Li-ion batteries is an obstacle to their extensive application. Improvements in the performance of supercapacitors are required to meet the needs of future systems, ranging from portable electronics to hybrid electric vehicles and large industrial equipment. Hence, the need for the development of new electrode materials and advances in our understanding of the electrochemical interfaces at the nanoscale level.

We, the editors, Subramanian, Sriraam, Nageswara Rao, and Aravinda Rao draw upon our extensive research experience and work in the field of energy storage to infuse this book with the required focus and perspectives. We would like to express our sincere thanks to our Management, viz., Ramaiah Institute of Technology, Bangalore, India (Shri M R Jayaram, Chairman, GEF, Shri M R Seetharam, Vice Chairman, GEF and Director, Ramaiah Institute of Technology, Dr N V R Naidu, Principal, Ramaiah Institute of Technology) as well as GITAM University, Reliance Industries Ltd, Vadodara, India for supporting this endeavour. Our thanks are also due to the distinguished authors who have contributed. We finally thank our family members for their support.

Professor K R V Subramanian
Professor N Sriraam
Professor T Nageswara Rao
Professor Aravinda C L Rao

Editor biographies

Prof K R V Subramanian

K R V Subramanian received his PhD from Cambridge University, Cambridge, UK, in 2006, specializing in nanotechnology. He is a Professor and HOD with the Mechanical Engineering Department and the Head of Research and Development at the Ramaiah Institute of Technology, Bengaluru, India. He has over 30 years of academic and industrial experience and has published over 150 journal and conference papers. He has been a principal investigator for many government-funded research projects

Dr N Sriraam

N Sriraam received his BE degree in Electronics and Communication Engineering (ECE) from the National Engineering College, Tamil Nadu, India, in 1996, his MTech degree (with distinction) in Biomedical Engineering from Manipal Institute of Technology (MIT), Manipal, India, in 2000, and his PhD in Information Technology from the Multimedia University, Cyberjaya, Malaysia, in the area of biomedical signal processing, in 2007. He is a Professor with Medical Electronics, Ramaiah Institute of Technology. His current research interests include biomedical signal processing, data mining, neural networks

Dr T Nageswara Rao

T Nageswara Rao received his doctorate in Mechanical Engineering from IIT Chennai, he has 31 years of experience (17 in teaching, 5 in research, and 9 in industry) in thermal engineering and computational fluid dynamics having particular expertise in the following areas: gas turbines (compressors, LPT and HPT, secondary flows, seals, TRF, TCF)—aerospace; data centers—electronic cooling; heat exchangers—process industries. He was Professor and HOD with Mechanical Engineering, GITAM University, Bengaluru

Dr Aravinda C L Rao

Dr Aravinda C L Rao works as a Assistant Vice President at Product Application and Research Center, Reliance Industries Ltd, Vadodara. After completing his doctoral studies at Bangalore University in 2001, he pursued his post-doctoral work at IISc, Bangalore, as an Alexander Humboldt fellow at Karlsruhe Institute of Technology, Technical University of Munich, Germany and University of California at Riverside, USA. His research interests encompass nanotechnology, graphene based advanced materials, polymer composites, biosensors, thin films and coatings for electronics. He has a rich and varied academic history as well as over 18 years of industrial R&D experience in multinational companies. He has authored more than 25 papers in reputed journals, a book chapter and several conference papers. He has handled large to medium scale R&D projects successfully and has established and managed large research labs.

List of contributors

H S Balasubramanya
Department of Mechanical Engineering, Ramaiah Institute of Technology, Bangalore, India

Partha Khanra
Chitkara University Institute of Engineering and Technology, Chitkara University, Punjab, India

Pankaj Kumar
Chitkara University Institute of Engineering and Technology, Chitkara University, Punjab, India

H Manjunatha
Department of Chemistry, Gitam School of Science, GITAM University, Bengaluru, Karnataka, India

Sudesh Mittal
Chitkara University Institute of Engineering and Technology, Chitkara University, Punjab, India

Niranjan Murthy
Department of Mechanical Engineering, Ramaiah Institute of Technology, Bangalore, India

B Ramakrishna Rao
Department of Chemistry, Gitam School of Science, GITAM University, Bengaluru, Karnataka, India

K Venkata Ratnam
Department of Chemistry, Gitam School of Science, GITAM University, Bengaluru, Karnataka, India

Venkatesh Sadhana
Department of Chemistry, Atria Institute of Technology, Bengaluru, Karnataka, India

Janardhan Sannapaneni
Department of Chemistry, Gitam School of Science, GITAM University, Bengaluru, Karnataka, India

Soorya Sasi
Advanced Molecular Materials Research Centre, Mahatma Gandhi University, Kottayam, Kerala, India

Gayatri Sharma
Chitkara University Institute of Engineering and Technology, Chitkara University, Punjab, India

K R V Subramanian
Department of Mechanical Engineering, Ramaiah Institute of Technology, Bangalore, India

Sunish K Sugunan
Department of Chemistry, CMS College Kottayam—affiliated to Mahatma Gandhi University, Kottayam, Kerala, India

IOP Publishing

Nanomaterials for Advanced Energy and Power Storage Devices

K R V Subramanian, N Sriraam, T Nageswara Rao and Aravinda C L Rao

Chapter 1

Nanomaterials for static and dynamic flow supercapacitors

Soorya Sasi and Sunish K Sugunan

The development of energy storage devices including batteries, fuel cells, and electrochemical capacitors is a Holy Grail of research in the energy sector. Batteries have low values of specific power and fuel cells are in the early stages of development. To remedy this bottleneck, supercapacitors, which show high specific power, power densities, and fast charging–discharging cycles, have recently been explored as a potential class of energy storage devices. In this chapter, we outline the development and applications of various nanomaterials in static and dynamic flow capacitors with emphasis on correlating their performance with their morphology, rheology, and their electrochemical properties.

1.1 Introduction

With the ever-increasing role of electrical and electronic devices in transforming the modern world into what it is today and what it could be in the future, steady supplies of energy are essential for the uninterrupted functioning of these devices [1]. This requires temporal storage of energy that can be supplied to devices in situations where the available energy is scarce or null to power them up. Energy storage devices have been developed to cope with this issue, the commonly used energy storage devices are batteries, fuel cells, and capacitors [2–4]. Batteries store energy by converting chemical energy into electrical energy through electrochemical discharge reactions. These devices can store a much greater amount of energy than any other storage device. Batteries release their stored energy on discharge by chemical reactions and inherit slight delays in power delivery, i.e., they are devices with high energy density, but with moderate power density. A fuel cell operates by converting the chemical potential energy stored in molecular bonds into electrical energy. These devices are high-energy systems, however, their capacity to deliver power is extremely poor. Electrolytic capacitors, on the other hand, are high-power

doi:10.1088/978-0-7503-4901-7ch1

delivery systems, though they deliver very small magnitudes of energy densities. A general observation can be made that electrolytic capacitors lack energy density, and fuel cells and batteries have lower power densities. Supercapacitors were invented to mitigate these issues, and these devices have progressed to a stage where they can deliver high power densities and reasonably good energy densities. It is worth noting that supercapacitors can not only be discharged in a matter of seconds but also be charged in a short period. This makes them attractive for applications that require rapid and short power bursts. A supercapacitor is also known as an ultracapacitor, and it stores charge in an electric double layer usually formed at the interface of a nanoparticle surface–electrolyte pair. The schematic structure of a supercapacitor depicting its working principle is given in figure 1.1.

The main components of a supercapacitor are two electrodes attached to the metallic current collectors, an electrode material coated onto the current collectors, and a separator inserted between the electrodes (the purpose of the separator is to avoid a short circuit) that electrically separates them, but allows passage of ions through them, enabling ion-transport between the electrodes. The final component is a suitable electrolyte. The electrodes are placed in an electrolyte solution with the ion-permeable separator at the center. The working of a supercapacitor is as follows: (1) when the electrode is dipped in the electrolyte, opposite charges are built spontaneously at the interface of the electrode and electrolyte, (2) the ions in the electrolyte are adsorbed onto the electrode surface, prompting the separation of charges between the electrode surface and the electrolyte, and (3) this leads to the formation of an electrochemical double layer at the interface of charge separation, and that can be used for charge storage. In a nutshell, when voltage is applied, the

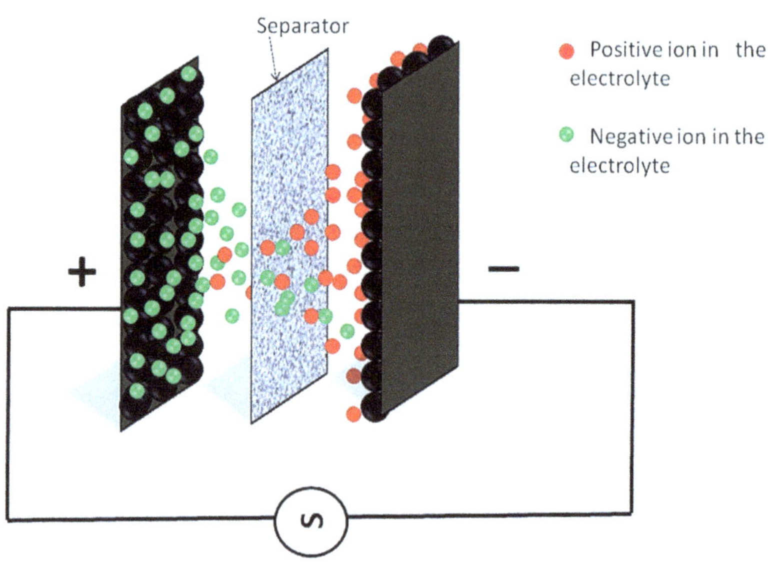

Figure 1.1. Schematic diagram of charge storage mechanism in a supercapacitor.

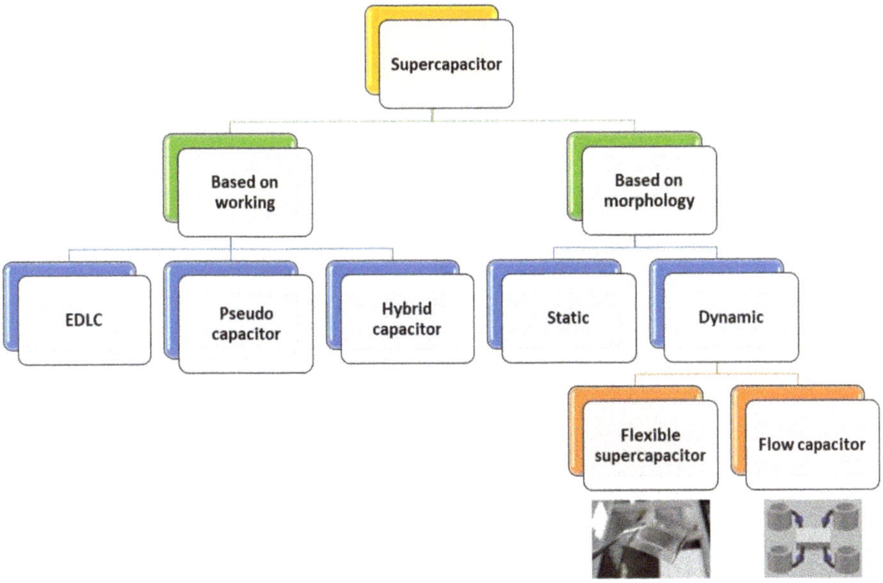

Figure 1.2. Flow chart showing the classification of supercapacitors.

electrode material on the positive electrode gets a positive charge, and that on the negative electrode gets a negative charge. Positive ions in the electrolyte are attracted toward the negative electrode, thus making a double layer where the charge is stored electrostatically. This happens at the positive electrode as well, where the negative ions from the electrolyte make the electric double layer. Supercapacitors can be classified into various types based on the operational energy storage mechanism and the morphology they possess and these are outlined in the flow chart (figure 1.2).

1.2 Classification based on the working principle

1.2.1 Electric double-layer capacitors

Electric double-layer capacitors (EDLCs) (figure 1.3) store charge electrostatically and are primarily composed of nanocarbon electrode materials coupled with suitable aqueous, ionic, or organic electrolytes. The capacitance in an EDLC is largely governed by the width of the binary layer at the electrode–electrolyte boundary, which is usually much smaller compared to the thickness of the separator.

The capacitance is measured in correspondence to the general capacitance equation,

$$\mathbf{C} = \frac{A \times \epsilon_0}{d}$$

where **C** denotes the capacitance measured in Farads, A denotes the surface area, ϵ_0 denotes the permittivity of free space and d denotes the effectual width of the electric

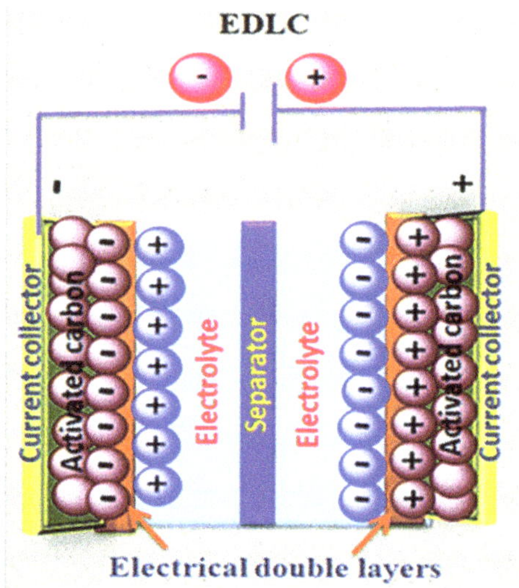

Figure 1.3. Schematic diagram of an EDLC. Reproduced from [31] with permission from the Royal Society of Chemistry.

double layer, termed the Debye length. The energy output of a normal EDLC depends on the electrostatic attraction among the ions at the electrode and the electrolyte in contact with the electrode.

1.2.2 Electrode materials of the EDLC

1.2.2.1 Porous carbon

To ensure enhanced performance, the pore size of the materials should be limited to the nanometer size regime [5]. Porous carbon-based electrodes are handy in this context as the small pore sizes of carbon materials render large specific surface areas to them. In addition, at a certain, optimal pore size, the effective surface area of the double layer can be maximized, concomitantly maintaining an optimal ion exchange rate with the electrolyte. To achieve precise control over the pore sizes in the carbon products, protocols were established by various groups for direct syntheses or seed growth of nanospheres of specific sizes.

1.2.2.2 Nanocarbon materials

Nanocarbon materials such as graphene, fullerene, carbon nanotubes (CNTs), and carbon nanofibers (CNFs) can improve the capacitive performance of an EDLC purely based on the principle of an electrochemical double layer [6]. Many high-surface-area mesopores in these types of nanocarbon materials promote ion-transport/charge storage. Graphene is a unique and attractive electrode material owing to its atom-thick two-dimensional (2D) structure and excellent electron transport and electrochemical properties. However, graphene sheets easily form irreversible agglomerates and restack into the graphitic structure. To overcome this problem, graphene can be hybridized or intercalated with CNTs, CNFs, and porous carbon.

1.2.2.3 Factors affecting the capacitance of an EDLC

Apart from the nature of the charge storage mechanism, the surface properties of the electrode material, its morphology, and the chemical structures of its components affect the electrochemical charge storage properties [7] of a supercapacitor.

- Specific surface area: for an electrode material that forms a film on the surface of the current collector, the specific surface area (a function of the abundance of the pores) of the film is directly correlated to the capacitance of the film electrode, as the enhanced surface area provides more active sites for accommodating the ions and promotes electrolyte diffusion.
- Pore-size distribution: pore size distribution has a crucial role in the capacitive properties because, to achieve excellent capacitor performance, a well-developed surface endowed with an optimal distribution of micropores that offer maximum accessibility to the electrolyte ions is required. A suitable amount of mesopores ensures rapid mass transport of ions within the electrode, facilitating quick charging and discharging of the double layer.
- Surface functionality: the capacitance depends on the wettability of the porous carbon electrode by the electrolyte, which ultimately is determined by the chemistry happening on the surface of the electrode. The presence of hydrophilic functional groups such as the -OH group on the carbon surface improves the wettability of the material by an aqueous electrolyte. Better wettability leads to better penetration of ions within the porous structure. In addition, some oxygen surface functionalities are known to induce pseudocapacitance which contributes to the enhancement of the overall capacitance.
- Pore size/shape and structure: the pore structures of porous carbon electrodes affect the redox characteristics of the redox-active materials adsorbed inside the pores. In addition, in cases where proton transfer or hopping mechanism is relevant in the supercapacitive performance, proton conduction usually depends on the pore structure of the electrode. It was found that carbon electrodes with mesoporous structures can facilitate better proton conduction.

1.2.2.4 Psuedocapacitors

When the pure carbon electrode material of an EDLC is replaced by a Faradaic material it is known as a pseudocapacitor [8]. Unlike the pure electrostatic storage mechanism operational in an EDLC, electrical energy is stored via electrochemical processes in a pseudocapacitor. These reactions are primarily redox in nature, initiated by the intercalation of specifically adsorbed ions on the electrode surface, followed by a change in the valence state of the electrode material because of the electron transfer. Pseudocapacitance arises due to the fast and highly reversible surface/near-surface redox reactions. Various charge storage mechanisms can be distinguished in a pseudocapacitor, such as underpotential deposition, transition metal oxide redox reactions, and intercalation pseudocapacitance. The Faradaic reaction occurs simultaneously with an EDL charge storage process in a pseudocapacitor, and this increases the specific capacitance of the device. The specific capacitance of a pseudocapacitor can be at least 100 times greater than that of an

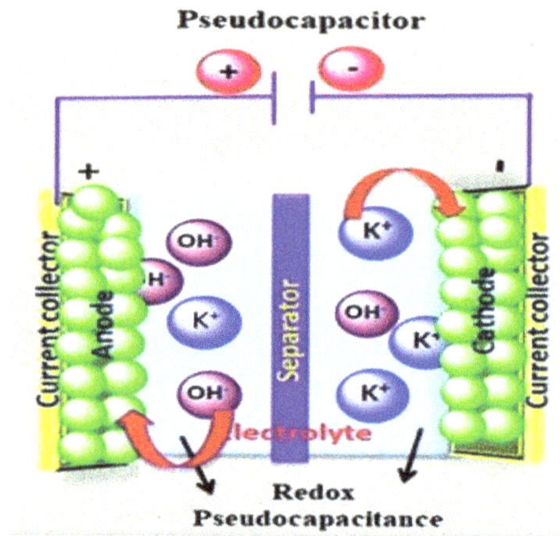

Figure 1.4. Schematic diagram of a pseudocapacitor. Reproduced from [31] with permission from the Royal Society of Chemistry.

EDLC. However, due to the slowness of the Faradaic processes involved in the charge storage mechanism, the power density of a pseudocapacitor is usually lower than that of EDLCs.

The pseudocapacitance (C) is generally represented as the derivative of charge acceptance (Δq) and changing potential (ΔV),

$$C = \mathrm{d}Q/\mathrm{d}V$$

A schematic diagram of the working principle of a pseudocapacitor is shown in figure 1.4.

1.3 Pseudocapacitive electrode materials

1.3.1 Conducting polymers

Conducting polymers [9] are a major category within pseudocapacitive materials. Conducting polymers such as polyaniline (PANI), polypyrrole (PPy), poly(3,4-ethylenedioxythiophene) (PEDOT), and their derivatives can undergo redox reactions efficiently, show relatively high specific capacitance and thus, are attractive candidates for pseudocapacitive applications. Among the conducting polymers, PANI is the polymer that has been most explored due to its ease of preparation, outstanding electrical conductivity, amenability to high levels of doping, high degree of environmental stability, and high specific capacitance values. Electrodes based on these conducting polymers usually show remarkably high specific capacitance and energy density compared to carbon-based electrodes, however, they deliver inferior power densities due to the sluggishness of the redox reactions happening at the interface of the conducting polymer and the electrolyte. Another drawback of conducting polymer-based supercapacitors is that they have poor cycling stability and relatively poor conductivity compared to carbon-based electrodes.

1.3.2 Metal oxides

Metal oxides [10] such as $Ni(OH)_2$, MnO_2, RuO_2, NiO, Co_3O_4, etc, are well-known pseudocapacitive materials. Among metal oxide pseudocapacitive materials, RuO_2 shows the best performance in terms of energy and power densities. But considering its availability and cost, other metal oxides are widely used. For the metal oxides, the Faradaic charge transfer process can be classified into three different types: (1) capacitive Faradaic storage, (2) intercalation pseudocapacitance, and (3) non-capacitive Faradaic storage. Capacitive Faradaic storage involves fast surface or near-surface electron-transfer reactions where the charge storage holds a linear relationship with the applied voltage within a specific electrochemical window. Faradaic charge transfer accompanied by no distinguishable crystallographic phase change is known as intercalation pseudocapacitance. The charge storage in some metal oxides such as NiO involves very fast surface reactions, but the amount of charge stored does not follow very linearly with the applied voltage within their working electrochemical window, and this is known as non-capacitive Faradaic storage.

1.3.3 Nanocarbon-metal oxide composites

Nanocarbon materials have excellent electrical and mechanical properties but exhibit very poor electrochemical properties due to their inert basal planes. On the other hand, metal oxides, such as RuO_2, $Ni(OH)_2$, MnO_2, NiO, etc, exhibit remarkable electrochemical properties and have proved their essential role in supercapacitor applications. These kinds of metal oxides commonly exhibit poor long-term cycling stability due to their relatively poor structural stability, active loss of materials, and over-oxidative decomposition. To solve this problem, nanocarbon-metal oxide composite electrodes have been employed [11]. Composite electrodes have several advantages, such as the synergistic combination of electronic and ionic conductivities, high surface area, controllable morphologies, high specific capacitance, and high energy density.

1.3.4 Hybrid supercapacitors

Hybrid supercapacitors [12] (figure 1.5) are combinations of EDLC and pseudocapacitors, with synergistic properties compared to the properties of their components. In an EDLC, the energy storage is based on intrinsic shell area and atomic charge partition length, whereas the energy storage is attained in a pseudocapacitor by the reversible redox reactions among active sites on active electrode materials and the electrolyte. The combination of storage mechanisms of both EDLC and pseudocapacitors together constitutes the energy storage mechanism of hybrid supercapacitors. One half of a hybrid supercapacitor acts as an EDLC while the other half behaves as a pseudocapacitor. Compared to their individual components, hybrid supercapacitors possess higher energy densities as well as power densities. This favors their use over other energy-storing devices in energy-efficient systems. Hybrid supercapacitors attempt to exploit the relative advantages and mitigate the relative disadvantages of EDLCs and pseudocapacitors to realize better performance

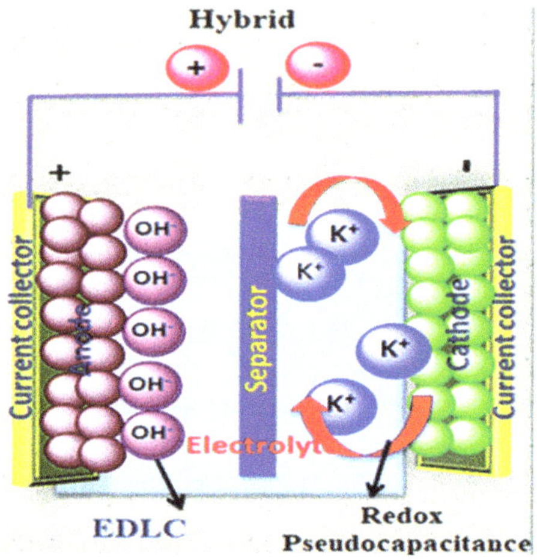

Figure 1.5. Schematic diagram of a hybrid capacitor. Reproduced from [31] with permission from the Royal Society of Chemistry.

characteristics. Utilizing both Faradaic and non-Faradaic processes to store charge, hybrid capacitors have achieved energy and power densities greater than EDLCs, albeit without sacrificing the cycling stability and affordability, which was the real issue that limited the success of pseudocapacitors.

Instead of using nanocarbon materials, metal oxides, and polymers separately, combinations of them can be used as hybrid electrode materials for supercapacitors, and this could provide the attributes of the device such as low cost, good electrical conductivity, mechanical flexibility, and good chemical stability. Hybrid electrode configurations usually consist of two different electrodes made of different materials. Composite electrodes consist of one type of material incorporated into another within the same electrode. Using pseudocapacitive materials capable of a high rate of intercalation for developing the anode is a promising way to balance the kinetics and power capability of both electrodes. The insertion-type materials are broadly classified into metal oxides, lithium/sodium metal oxide-based composites, transition metal carbides, and transition metal dichalcogenides, etc. However, of these materials, lithium, and sodium metal-based materials and rutile TiO_2 cannot be defined as pseudocapacitive materials because they show plateauing of voltage and undergo phase changes during the deintercalation process. But their high charging and discharging rate ensures their applicability as high-rate electrodes in hybrid capacitors.

Hybrid capacitors can be classified into three different types, distinguished by their electrode configuration:

- asymmetric hybrids
- battery-type hybrids
- composite hybrids

Asymmetric hybrids combine Faradaic and non-Faradaic processes by coupling an EDLC electrode with a pseudocapacitive electrode whereas battery-type hybrids

Table 1.1. Comparison between EDLC, pseudocapacitors, and hybrid capacitors.

Parameter	EDLC	Pseudocapacitors	Hybrid capacitors
Energy storage	Electrostatic	Faradaic	Both electrostatic and Faradaic
Energy density	Low	High	High
Power density	High	Low	Intermediate
Specific capacitance	Low	High	High
Cycling stability	High	Low	Intermediate
Commonly used materials	Carbon and nanocarbon materials like activated carbon, graphene, CNT, fullerene, etc	Metal oxides/hydroxides (NiO, $Ni(OH)_2$, RuO_2, MnO_2, TiO_2, etc) Polyoxometalates/metal sulfides (Polymolybdate, Niobates, Vanadates, MoS_2, PbS, Conducting polymers like PANI, Polypyrrole, PEDOT, etc)	Mixture/composites of commonly used pseudocapacitive materials

couple two different electrodes, i.e. a combination of a supercapacitor electrode with a battery electrode. Composite hybrids integrate carbon-based materials with either a conducting polymer or metal oxide materials and incorporate both physical and chemical charge storage mechanisms together in a single electrode.

Table 1.1 provides a comparison between EDLC, pseudocapacitors, and hybrid capacitors.

1.3.5 Classification based on the morphology

Much research is being carried out to develop stable and good-performing supercapacitors. A notable achievement in this direction has been achieved by the researchers at Central Florida University; they developed a paradigm-shifting supercapacitor battery that is much smaller than the lithium-ion cells. This device has a very fast charging time and demonstrates a cycling life of 30 000 charging and discharging cycles with a negligible decrease in performance, demonstrating significant advances in the state-of-the-art field. Another ground-breaking invention is the recent development of lightweight graphene-based supercapacitors with exceptional energy storage capabilities as high as 550 $F\ g^{-1}$ at a remarkably low cost, i.e., at a fraction of the price of current EDLC designs. A proper material selection must be made considering the requirements of the final application such as long life span, energy density, and power density. Apart from the materials, the design, development, and optimization of cell configurations represent a growing field of opportunities for developing hybrid battery/

supercapacitor systems. Such systems will be in great demand in those applications where a battery or supercapacitor alone does not meet specific needs such as energy density, cycle life, and power rating. Based on the morphology requirements, supercapacitors can be subdivided into static and dynamic supercapacitors.

1.3.6 Static/rigid devices

Static/rigid devices (figure 1.6) are inflexible but are very popular, largely due to their low cost of production. Flexible alternatives, however, are beginning to take over the market from rigid ones due to their versatility.

1.4 Dynamic devices

1.4.1 Flexible supercapacitors

Dynamic supercapacitors can be subdivided into flexible supercapacitors and flow capacitors. Figure 1.7. shows photographs of flexible supercapacitors.

Flexible devices can have the following advantages [13]:

- **Flexibility.** Flexible supercapacitors can be bent, folded, and even creased to fit the end application, giving the designer the ability to easily fit the circuitry into the device. This is advantageous as conventional devices are made in such a way that the end device has to be built around the electronics and circuit boards, which really constrains the manufacturers' ability to mold the device into desirable shapes. This makes flexible supercapacitors ideal for wearable electronics applications.
- **Connectivity.** Flexible supercapacitors provide greater connectivity with other elements in the circuit boards, electronic components, and the user interface

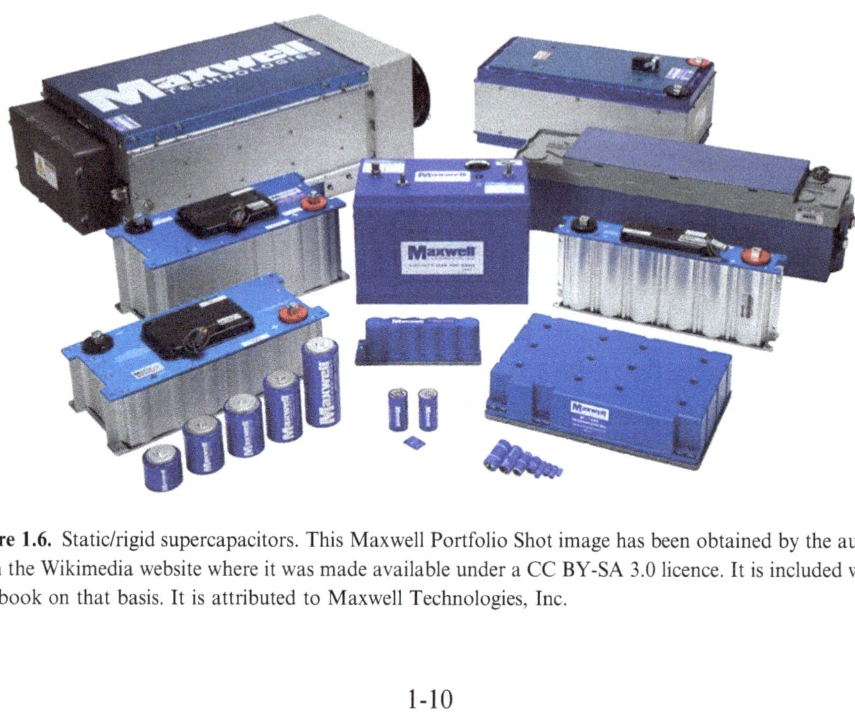

Figure 1.6. Static/rigid supercapacitors. This Maxwell Portfolio Shot image has been obtained by the authors from the Wikimedia website where it was made available under a CC BY-SA 3.0 licence. It is included within this book on that basis. It is attributed to Maxwell Technologies, Inc.

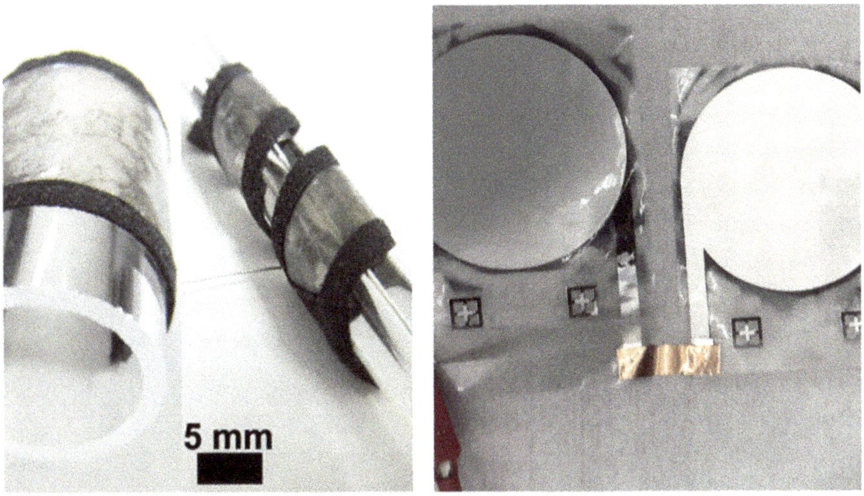

Figure 1.7. Photographs of flexible supercapacitors—left panel: flexible supercapacitor based on carbon nanocups, and right panel: two flexible supercapacitors made from based on cellulose nanofibrils (CNF) and the conducting polymer PEDOT:PSS and connected in series (left panel: reprinted from [32], copyright (2012), with permission from Springer; right panel: reproduced from [33]. CC BY 4.0.

in electronic packaging. They can even provide connectivity in dynamic flexible applications where the flexible circuit may be flexed as and when required over the life of the device so that they can be extensively used in laptop computers, foldable electronics, and display applications.

- **Reduced weight.** A low-weight circuit board results in a light end product, which is essential in today's electronics market, where compact and light-weight devices are preferred by electronic device designers and consumers. Flexible supercapacitors are ideal for flexible/portable/wearable electronics due to their very light weight.
- **Durability.** Flexible polymer-based devices absorb shocks and vibrations much more effectively than their rigid counterparts. This provides long-term reliability, long life, and functionality.

Flexible energy storage devices are essential to the development of next-generation wearable, compact, and portable electronics, e.g., flexible displays on health tracking devices, phones, computers, and televisions. With the recent boost of consumer electronics, society requires supercapacitors to be foldable/flexible. With increased market needs, nowadays, meticulous attention is being dedicated to developing flexible supercapacitors. To this end, flexible supercapacitors are highly attractive in comparison to batteries (LIBs) as they combine the inherent high-power density (>10 kW kg^{-1}), fast charging/discharging capability, longer operation lifetime, and mechanical flexibility. Flexible and/or transparent supercapacitors are certainly some of the best solutions to meet the specific needs of the future electronic world. There is an urgent need for the time to develop supercapacitors with mechanical flexibility. Electrodes typically used in flexible supercapacitors are

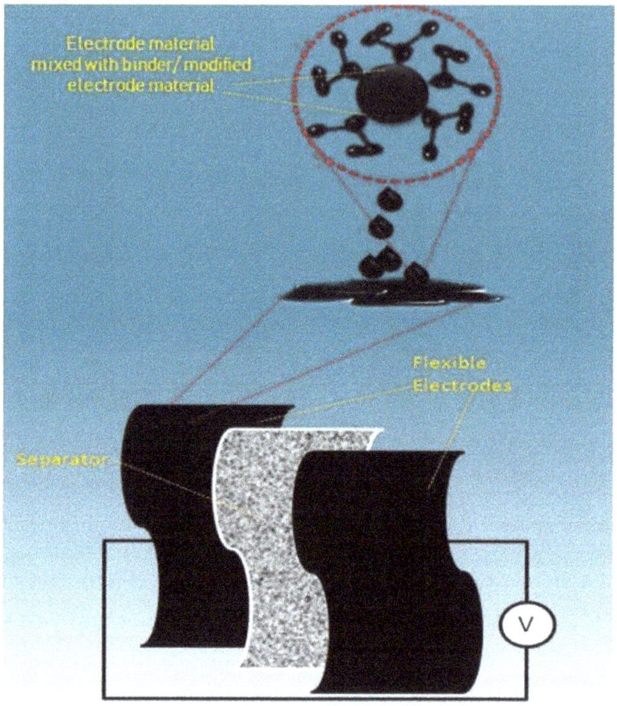

Figure 1.8. Schematic diagram of a flexible supercapacitor.

usually made of carbon-based materials as the basic unit. In flexible supercapacitors, the highly conducting and flexible substrate such as a carbon network serves as both the electrode and current collector. In contrast, conventional supercapacitors consist of an outer case, current collectors in the form of metal foils, and positive and negative electrodes in electrolytes separated by an ion-transport layer. As such, flexible supercapacitors with carbonaceous materials as electrodes are lighter and easier to be constructed than traditional, non-carbonaceous electrode-based supercapacitors. Another key difference from conventional supercapacitors is that each component in flexible supercapacitors (e.g., electrodes, separator, and outer packing shell) can be made flexible. To enhance the performance of pure carbon-based supercapacitors, metal oxides (such as RuO_2, MnO_2, In_2O_3, CuO, Co_2O_3, and Ni $(OH)_2$ or NiO) and conducting polymers (such as polypyrrole, PEDOT or polyaniline) have been incorporated into the carbon network of the electrode. A mixture of the carbon network with the pseudocapacitive materials enhances the performance of flexible electrodes (*vide supra*). A schematic and an example of a flexible supercapacitor is shown in figure 1.8.

1.5 Materials used as substrates in flexible supercapacitors

Metal substrates are commonly chosen as electrode substrates [14] for flexible supercapacitors due to their high electrical conductivity and favorable mechanical properties. Examples of such metal substrates include copper sheets, nickel sheets, and titanium foams, all of which offer excellent strength, conductivity, and ease of preparation. Additionally, configuring electrodes with electroactive materials directly synthesized on the metal substrate surface can potentially enhance both energy density and flexibility [15]. In contrast, conventional materials often necessitate the use of additional binders, leading to lower energy density and mechanical stability. Furthermore, flexible electrodes supported by metal substrates tend to be opaque and less flexible, highlighting the need for flexible plastic substrates in supercapacitor electrodes. Flexible supercapacitor electrodes incorporating graphene films on substrates like polyethylene terephthalate (PET), indium tin oxide-PET, or single-walled carbon nanotubes on polydimethylsiloxane and polyaniline substrates have shown significant promise in transparent electronics due to their excellent capacitance and stretchability [16]. However, plastic substrates face challenges regarding limited electrical conductivity.

Porous metallic substrates are often favored for flexible supercapacitor applications due to their significant conductivity and porosity. However, their limited corrosion resistance restricts their suitability. Additionally, the heavy nature of metallic current collectors commonly used further adds to the device's weight, impacting portability.

In contrast, carbon-based supporting substrates like carbon paper, foam, and cloth offer superior electrical conductivity, reduced corrosive susceptibility, enhanced flexibility, and lower weight compared to their metallic counterparts. These attributes make them more suitable for flexible supercapacitor applications [17]. Flexible supercapacitors employing paper as a substrate boast attributes such as being lightweight, bendable, transparent, and easy to process, making them attractive alternatives for electronic screens in portable devices. Notably, research into free-standing electrodes composed of carbon nanotube(CNT)–paper composites and the deposition of pseudocapacitive materials on recyclable paper substrates has yielded positive results. The macroporous and network-free structure of flexible sponge substrates facilitates high adsorption, increased surface area, uniform coating, and improved interaction between electrodes and electrolyte, leading to enhanced performance. Symmetrical flexible supercapacitors incorporating a CNT sponge assembly, synthesized via chemical vapor deposition, have demonstrated excellent cyclic stability.

When evaluating flexibility and stretchability, textile-based substrates [18] offer several advantages over paper-based alternatives. The porous nature of textile substrates provides ample support for active material loading and promotes rapid adsorption of electroactive substances due to their hydrophilic properties. This leads to a significantly higher areal mass loading of active materials, resulting in superior areal power and energy density in flexible supercapacitors. Textile substrates such as cotton cloth, polyester microfiber twill, and carbon fabric are typically produced from natural or synthetic fibers using methods like weaving, pressing, knitting, or felting. These substrates offer several advantages for flexible supercapacitors, including high stretchability, lightweight nature, a three-dimensional open-pore structure, good mechanical strength, and cost-effectiveness compared to other substrate options. Stretchable textile electrodes are commonly created by incorporating single-walled

Table 1.2. Various types of substrate materials used in flexible supercapacitor devices.

Carbon materials	Metals	Polymers	Other materials
Carbon cloth	Ni foam	PET	Air-laid paper
CNT/CNF	Cu foam	PDMS	mCel-membrane
Carbon paper	Stainless steel fabric	Au coated PET	Fiber/yarn
Graphene	Ti wire, Au wire	Polydimethylsiloxane, ethylene/vinyl acetate polycarbonates, polyethersulfone, polyimide	Reduced graphene (rGO) coated conducting yarn

carbon nanotubes into cotton cloth or porous carbon materials into woven cotton/polyester textiles. Carbon cloth, known for its robustness, stiffness, and flexibility, is often manufactured using weaving techniques such as plain, satin, or twill weaving. To initiate the formation of various carbon networks, a stable dispersion of carbon material combined with a ligand like sodium dodecylbenzene sulfonate in a suitable solvent serves as a crucial starting agent. This method is utilized for producing different carbon structures, including carbon films, papers, and textiles, all of which find applications in flexible supercapacitors.

Inexpensive and lightweight plastic substrates like polyethylene terephthalate (PET), polydimethylsiloxane (PDMS), and ethylene/vinyl acetate copolymer films show promise as supportive materials for active components in flexible supercapacitors. PET is particularly favored due to its widespread availability, excellent resistance to water and moisture, and transparency. Carbon films can be manufactured using chemical vapor deposition (CVD) techniques, allowing for the creation of single-walled carbon nanotubes on substrates resembling polydimethylsiloxane. Alternatively, carbon films can be produced through ink-jet printing or spin-coating carbon materials onto flexible plastic or paper substrates. Paper, with its high porosity and surface area, is a highly suitable substrate. However, the presence of large pores can lead to carbon nanostructures penetrating the substrate. To address this, a polyvinylidene fluoride coating can be applied as a straightforward pre-treatment method. This coating enhances the adhesion of carbon electrode materials to the porous paper substrate while preserving electrical conductivity.

Various other flexible polymer substrates have been tested for use in flexible supercapacitors, including polycarbonates, polyethersulfone, and parylene. These substrates offer additional options for designing and fabricating efficient and durable flexible supercapacitor devices. Table 1.2 shows the list of substrate materials used in flexible supercapacitors.

1.6 Materials used for making the electrode

Recent research has explored a diverse range of new pseudocapacitive electrode materials aimed at enhancing the energy density of flexible supercapacitors. Materials such as NiO, RuO_2, Fe_3O_4, and MnO_2 have attracted attention due to their ability to increase the energy density of carbonaceous sponge substrates [19]. In efforts to further boost electrochemical capacitance, metal oxide nanoparticles have been applied to carbon fiber-based textile substrates, forming flexible electrodes. The coating of electrochemically active materials plays a crucial role in enhancing the total capacitance of textile-based flexible electrodes, as textiles inherently possess low capacitance. Various approaches have been employed, including the fabrication of free-standing electrodes incorporating metal oxides/nitrides, polymer/CNT (carbon nanotube), and graphene hybrids. These methods leverage synergistic effects from the exceptional electrochemical performance of pseudocapacitive materials and the high conductivity and mechanical robustness offered by CNTs and graphene. These advances in electrode material development and fabrication techniques hold significant promise for achieving higher energy density and improved performance in flexible supercapacitors.

Additionally, metal–organic frameworks (MOFs), a class of coordination polymers having high specific surface area and manageable pore size to anchor the active materials, are being used to address the drawbacks of carbon textile substrates [20]. MOFs can be applied in three different ways for developing flexible supercapacitors: directly as an electrode material, as a composite electrode, or as a flexible substrate for active electrode materials. Polyoxometalates are another class of porous substrate with novel electronic properties, robust structure, and the capability to behave like an acid during synthesis owing to their metal-oxygen clusters. They have recently been used to substantially enhance the electrical conductivity of low-cost electrode materials like MnO_2 nanoparticles. Another developing category in flexible substrates is biomass-derived substrates and wood substrates. Wood transverse section slice is a promising candidate for flexible substrates as it shows excellent hydrophilic properties but does not require the harsh or expensive chemical processing steps involved in the preparation of other substrates (e.g., nanocellulose paper).

Graphene has excellent optical, electrical, and mechanical properties, has important application prospects in energy, micro-nano processing, materials science, biomedicine, and drug delivery, and will be a revolutionary material in the future. Like activated carbon, graphene also has a high specific surface area, high conductivity, and excellent electrochemical activity. Graphene has been the most widely used flexible supercapacitor electrode material [21]. Flexible electrodes can be fabricated on textiles by screen printing with graphene oxide ink, and the graphene oxide is reduced *in situ* by a rapid electrochemical method. Due to the strong interaction between the ink and the textile substrate, the electrodes exhibit excellent mechanical stability. The CNTs are usually multi-wall structures, which consist of several hundred to tens of layers of coaxial tubes composed of hexagonally arranged carbon atoms. CNTs have many excellent properties such as heat resistance, corrosion resistance, and thermal conductivity. In addition, CNTs have excellent mechanical, electrical, optical, and chemical properties with unique structural characteristics. Meanwhile, CNTs have a large specific surface area, high electrical conductivity, and many chemical reaction sites. Therefore, CNTs are being studied for many applications in electrode materials for supercapacitors. Flexible supercapacitors can be fabricated by printing CNTs inks on different substrates through screen printing and spray printing. Compared with

Table 1.3. Materials used for making the electrode

Polyoxometalates/ metal sulfides	Metal oxides	Metals and QDs	Carbon materials	Other materials
Polymolybdate	SiO_2	Au	CNT	PANI
Niobates	TiO_2	Pt	Fullerene	Anthracene
Vanadates	ZnO	Pd	Graphene	Calcium carbonate
MoS_2	Fe_2O_3	Ag	Carbon black	Carbazole
PbS	$ZrO_2/Ni(OH)_2$	CdSe/CdTe quantum dots	Reduced graphene cloth	Porphyrins
ZnS, SnS, Cu_7S_4	MnO_2	Ni	Activated carbon/ activated carbon cloth	Polyaniline hydrogel

other carbon materials, CNTs are excellent electrode materials for flexible supercapacitors, owing to their high electrical conductivity and controllable regular pore structure [22]. Moreover, CNTs possess a high aspect ratio, which provides long continuous conductive paths and ensures high flexibility. Vacuum filtration is among the most widely used methods in CNT film preparation and can be easily achieved in a laboratory with simple equipment. However, the area of CNT films prepared through this method is limited by the filter size.

Another category of flexible supercapacitor electrodes is polymer gel-based electrodes [23]. These solid-state supercapacitors are typically assembled by sandwiching a polymer gel electrolyte between two electrode pellets or films (active materials deposited on metal substrates, such as stainless steel, and pressed together into one device). This approach of first fabricating the electrodes and polymer electrolyte separately and subsequently assembling them together have two major shortcomings: firstly, the general use of thick solid-state gel electrolytes and metallic current collectors limits any attempts to reduce the device thickness beyond a limit. Thus, the middle polymer gel electrolyte needs to be as thin as possible but thick enough to separate the two electrodes. Secondly, because the solid electrodes are in contact with the solid-state electrolyte under pressure, only the electrode part near the geometric electrode/electrolyte interface can be utilized effectively. Table 1.3 lists the materials used for making the electrode.

1.7 Materials used as the electrolyte

Electrolytes are key components in flexible supercapacitors and the selection of electrolyte significantly affects the electrochemical properties of the whole device, such as energy density, rate capability, and cycling stability. Usually, when a flexible supercapacitor is fabricated, either a liquid electrolyte is infused in the electrode/ separator layer or a solid-state electrolyte is sandwiched between the electrodes [24]. Solid-state electrolytes act as both the electrolyte as well as the separator by playing a dual role in the device, such as ionic conducting media and electrode separators.

They are widely used to fabricate flexible supercapacitors since they provide good mechanical stability, which allows the successful assembly of flexible supercapacitors in various applications. The key to developing a high-performance device is a suitable solid-state electrolyte that exhibits high ionic conductivity; high chemical, electrochemical and thermal stabilities; good mechanical strength; and dimensional stability.

Three types of solid-state electrolytes are mainly used in flexible supercapacitors: ceramic electrolytes, gel polymer electrolytes, and polyelectrolytes. Among these, gel polymer electrolytes are widely used in flexible supercapacitors due to their relatively high ionic conductivity. Normally, gel polymer electrolytes are composed of a polymeric host such as PVA/PVP/PVDF, a solvent as the plasticizer, and a conducting electrolytic salt. The polymer serves as a medium through which the solvent and the ions travel through. Gel polymer electrolytes are subdivided into the following categories.

The first category is aqueous gel polymer electrolytes which contain a host polymer matrix (PVA/PVP/PMMS), water as a plasticizer, and an electrolytic salt that can be a strong acid (H_2SO_4/H_3PO_3), strong alkali (KOH), or neutral salt (LiCl, Na_2SO_4). They are also known as hydrogel polymer electrolytes, in which 3D polymeric networks trap water molecules through surface tension. The next category is organic gel electrolytes with a narrow operating voltage window which limits their wide application. Generally, organic gel electrolytes consist of a physical blend of a high molecular weight polymer (PMMA/PVDF–HFP) gelated with a conducting salt dissolved in a non-aqueous solvent. Organic gel electrolytes show better ionic conductivity. Organic solvents, such as propylene carbonate, ethylene carbonate, dimethyl formamide, and their mixtures have been used as common plasticizers.

Then comes Ionic liquid-based gel polymer electrolytes which exhibit some additional advantages over aqueous and organic electrolytes, including high ionic conductivities, non-volatility, non-flammability, and wide potential window. Therefore, they are considered suitable for use in flexible supercapacitors. Like the other two categories, various polymer hosts including VDF–HFP/PVA and PEO are used in Ionic liquid-based gel polymer electrolytes also. Ionic liquids such as 1-butyl-3-methylimidazolium bis(trifluoromethyl sulfonyl)imide/1-butyl-3-methylimidazolium chloride/1-ethyl3-methylimidazolium tetrafluoroborate/*N*,*N*-dimethyl acrylamide are used as supporting electrolytes.

The final category is redox couple electrolytes. An evolving strategy to improve the electrochemical performance of a device is to modify/add redox-active species to the electrolyte to maximize the capacitance and consequently the energy density of the device. The redox couple used as the electrolyte is usually made by mixing parts of a chemical species existing in different oxidation states that can add extra pseudocapacitance through reversible Faradaic reactions and fast electron transfer at the electrode–electrolyte interface, resulting in significant enhancement of the specific capacitance. Iodide/triiodide (I^-/I_3^-) based electrolytes where (I^-/I_3^-) redox couple dissolved in a suitable organic solvent such as nitriles including propionitrile, valeronitrile, glutaronitrile, methoxyacetonitrile, etc, are typical examples for redox couple electrolytes. The iodide can produce redox pairs (I/I_3 and I_2/IO_3) during the

Table 1.4. Components of solid-state electrolytes used in the flexible supercapacitors

Component	Aqueous gel polymer electrolytes	Organic gel electrolytes	Ionic Liquid-based gel polymer electrolytes	Redox couple electrolytes
Polymer host	PVA, PVP, PVDF, PAM, PEO, PAA etc	PAN-b-PEG-b-PAN, PVDF–HFP, PEO, etc	PEGDA, PVA, PVDF–HFP, PEO, PEOEMA, PHEMA/chitosan, etc	PVA, PMMA etc
Plasticizer	Water	Organic electrolyte	Ionic liquid	Water/organic/ionic liquid
Electrolytic salt	H_2SO_4, LiCl, SiWA–H_3PO_4, BWA, KOH, TEAOH, SiO_2–SiWA–H_3PO_4, graphene oxide doped KOH	DMF–$LiClO_4$, PC–NaTFSI, PC–EC-DMC–NaTFSI, PC–$Mg(ClO_4)_2$, PC–EC–$LiClO_4$, PC–EC–$NaClO_4$	$BMIPF_6$–$LiPF_6$, EMIMCl, EMITf, EMITf–NH_4Tf, [BMIM] [TFSI], [EMIM] [TFSI], $EMIHSO_4$–$ImHSO_4$, $EMIHSO_4$–$MIHSO_4$	BAAS doped H_2SO_4, AQQS doped H_2SO_4, p-Benzenediol doped H_2SO_4, PC-Fc doped TEABF, PC-4-oxo TEMPO doped $TEABF_4$

electrochemical process. The ionic sizes of I, and I_3 are small so they can easily access the micropores and small mesopores of the porous electrode. Numerous redox couples, such as iodides (KI), $K_3Fe(CN)_6$, and Na_2MoO_4, organic redox mediators, such as hydroquinone, p-phenylenediamine, p-benzenediol, methylene blue anthraquinone, 2,7-disulfonate, indigo carmine, etc, have been investigated in solid-state gel electrolytes. This approach of adding a redox couple of electrolytes effectively improves the performance of the device. Table 1.4 lists the components of the solid-state electrolytes used in the flexible supercapacitors.

1.8 Electrochemical flow capacitor

Another category belonging to the dynamic supercapacitors group is the electrochemical flow capacitor (EFC), which is a new technology for grid energy storage and is based on the fundamental principles of supercapacitors [25]. The EFC concept benefits from the advantages of both supercapacitors and flows batteries in that it is capable of rapid charging/discharging, has a long cycle lifetime, and enables energy storage and power to be decoupled and optimized for the desired application.

Fundamentally, the EFC is based on the same charge storage mechanism as supercapacitors, where reversible polarization leads to the formation of the electric double layer by counterbalancing the surface charges of porous electrodes. While it uses a flow cell architecture, like existing redox flow batteries for grid storage,

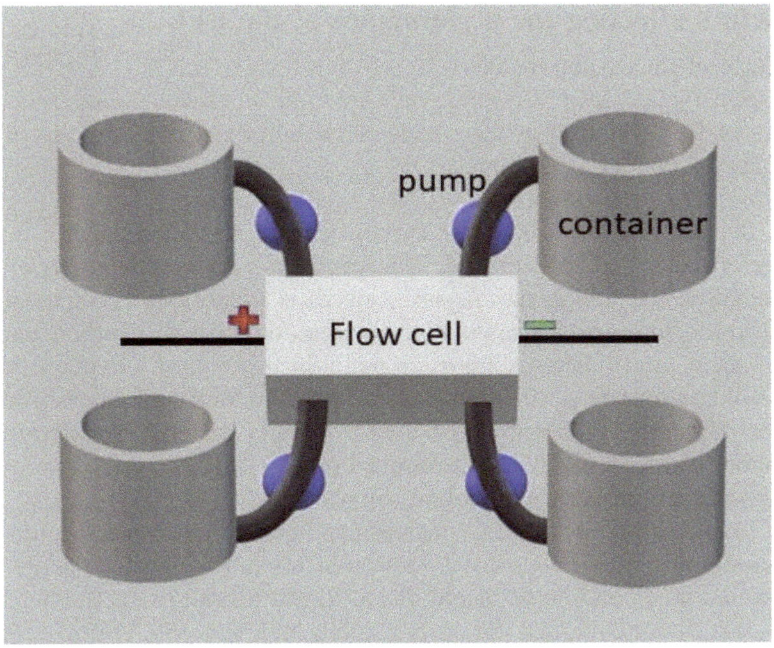

Figure 1.9. Schematic diagram of an electrochemical flow capacitor.

consisting of an electrochemical cell connected to external electrolyte reservoirs. However, this technology is exceptional in that it uses a flowable slurry of electrochemical material particles suspended in a liquid electrolyte carrier fluid. The operation of an EFC occurs in a way that an uncharged slurry (usually a flowable mixture of porous carbon material with aqueous or organic electrolyte) is pumped through a polarized flow cell, where energy storage occurs through the formation of the electric double layer at the surface of the carbon particles. Containers outside the flow cell are filled with a carbon-based slurry (carbon + electrolyte). On charging, it is pumped into the flow cell, and the charge is stored in the carbon particles electrostatically [26]. Positively charged carbon particles in the slurry attract negatively charged ions for charge balancing. On discharging, negatively charged carbon particles in the slurry attract positively charged ions. Once charged, the slurry is pumped into an external reservoir for storage and when it is necessary to recover the stored energy, the charged slurry is again pumped through a flow cell. Charging can occur very rapidly, yet power output and energy storage are decoupled, which enables energy capacity, overcoming the major limitation of supercapacitors, i.e., the moderate amount of stored energy. Figure 1.9 shows the schematic diagram of an EFC.

Carbon materials like phenolic resin-derived activated carbon beads, non-spherical carbide-derived carbon powder, and carbon black-added activated carbon are commonly used as active materials in EFCs [27]. The well-known nanocarbon forms, such as carbon nanotubes, graphene, and fullerenes, have demonstrated their efficiency as electrode materials for EFC supercapacitors.

1.9 Factors affecting the performance of the EFC

- Shape of the carbon material

  Spherical-shaped slurry materials are more suitable for suspension electrodes in EFC because of their better flowability compared to that irregular-shaped active material powder.
- Size of the carbon material

  By reducing the size of the carbon spheres to the nanometer range, the performance can be improved. The specific surface area increases as the size decreases and a high specific surface area is a desirable characteristic of an electrode material for any energy storage device [28]. Carbon nanotubes, graphene, and fullerenes are quite useful in this regard. Graphene is a good candidate for use as the carbonaceous slurry material of an electrochemical flow capacitor. Fullerene is also used as a slurry material considering its size, shape, and substantial electrochemical properties.
- Rheological properties of carbon slurry

  For a slurry in an EFC, increasing the solid-component content while maintaining a low viscosity is important for improving the energy density. Beyond a percolation threshold, the solid component in the slurry constructs an infinite network, and the slurry exhibits high viscosity. Porous carbon with a high surface area and large pore volume adsorbs much more electrolyte in the pore, resulting in a higher viscosity of its slurry. So, filling the pores of the carbon beads with organic compounds reduces the amount of electrolyte inside the carbon beads, increasing the apparent volume of the solution outside them leading to a lower viscosity.
- Introducing pseudocapacitive materials

  With the use of pseudocapacitive materials such as metal oxides, hydroquinone, and p-phenylenediamine, specific capacitance can be increased [29]. If nanocarbon materials are used instead of micronized carbon particles for making the capacitive slurry, EFC can be used in high-power applications for hundreds of thousands of charge–discharge cycles and could become vital for industrial applications. The charge storage capacity of the conventional carbon slurry can be improved by complexing it with pseudocapacitive materials like redox-active quinone derivatives (a proton–shuttle system between the electrolytes during the charge/discharge operations) to address the energy density limitations of EFC slurries and could potentially move the technology towards large-scale energy storage in smart grids. To avoid the reduction in the slurry flow rate due to the irregular shapes of the pseudocapacitive material introduced to the carbon slurry, the pseudocapacitive material can be enveloped in the carbon beads.
- Engineering design of the flow cell

  The engineering parameters like the feed flow rate and solid concentration, the rheology properties of suspension electrodes, and flow cell design, can be optimized to improve not only the electrochemical performance but also the long-term stability of suspension electrodes [30]. In the electrochemical flow capacitor, the energy storage capacity is determined by the size of the reservoirs which store the charged material. If a larger capacity is desired, the tanks can simply be scaled up in size. Similarly, the power output of the system is controlled by the size of the electrochemical cell, with larger cells producing more power. The design of the flow capacitor cell and the flow channel also play crucial roles in the performance of the EFC. To minimize ohmic losses, it is generally advantageous to use a very shallow channel geometry, which allows free passage of the slurry without clogging. The overall length of the channel determines the charging rate and the residence time of the slurry within the cell. In addition, it is possible to lower the contact resistance and achieve higher flow rate capability through further engineering of the current collectors. The performance of an EFC can be improved by expanding the contact area of current collectors exposed to the flowing slurry.

1.10 Conclusion

Supercapacitors are one of the most important inventions in the field of energy storage devices. They are of mainly three categories: EDLC, pseudocapacitors, and hybrid capacitors. Based on their morphology they can be categorized into static and dynamic devices. Dynamic devices are mainly of two types: flexible and flow capacitors. A detailed account of the various electrochemical materials that have been explored to develop the substrate, electrode, and electrolyte components of these devices is given in this chapter. Among these materials, pseudocapacitive materials are found to enhance the energy density of the supercapacitors. We attempted to draw a clear picture of the dependence of the performance of the device with its components and constituent materials. The content covered in this chapter indicates that judicious selection of the active material, clever choice of the electrolyte and proper design of the device – all these factors are of pivotal importance in deciding the performance of a supercapacitor device.

References

[1] [a] Finley J W and Seiber J N 2014 The nexus of food, energy, and water *J. Agric. Food Chem.* **62** 6255–62

[b] Blazev A S 2021 *Global Energy Market Trends* (Boca Raton, FL: CRC Press)

[c] Armaroli N and Balzani V 2007 The future of energy supply:challenges and opportunities *Angew Chem. Int. Ed.* **46** 52–66

[2] [a] Chen H, Cong T N, Yang W, Tan C, Li Y and Ding Y 2009 Progress in electrical energy storage system: a critical review *Prog. Nat. Sci.* **19** 291–312

[b] Hadjipaschalis I, Poullikkas A and Efthimiou V 2009 Overview of current and future energy storage technologies for electric power applications *Renew. Sustain. Energy Rev.* **13** 1513–22

[3] [a] Kramer D, Lehmann E, Frei G, Vontobel P, Wokaun A and Scherer G G 2005 An Online study of fuel cell behavior by thermal neutrons *Nucl. Instr. Meth. Phys. Res Sec.* A **542** 52–60

[b] Ajanovic A and Haas R 2021 Prospects and impediments for hydrogen and fuel cell vehicles in the transport sector *Int. J. Hydrogen Energy* **46** 10049–58

[4] [a] Sharma P and Bhatti T S 2010 A review on electrochemical double-layer capacitors *Energy Convers. Manag.* **51** 2901–12
[b] Wei B and Ajayan P M 2011 Direct laser writing of micro-supercapacitors on hydrated graphite oxide films *Nat. Nanotechnol.* **6** 496–500
[c] Yuan L *et al* 2012 Flexible solid-state supercapacitors based on carbon nanoparticles/ MnO_2 nanorods hybrid structure *ACS Nano* **6** 656–61
[5] [a] Choi H and Yoon H 2015 Nanostructured electrode materials for electrochemical capacitor applications *Nanomaterials* **5** 906–36
[b] Luo X Y, Chen Y and Mo Y 2021 A review of charge storage in porous carbon-based supercapacitors *New Carbon Mater.* **36** 49–68
[6] Mirzaeian M, Abbas Q, Ogwu A, Hall P, Goldin M, Mirzaeian M and Jirandehi H F 2017 Electrode and electrolyte materials for electrochemical capacitors *Int. J. Hydrogen Energy* **42** 25565–87
[7] [a] Zhang S and Pan N 2015 Supercapacitors performance evaluation *Adv. Energy Mater.* **5** 1401401
[b] Luo X Y, Chen Y and Mo Y 2021 A review of charge storage in porous carbon-based supercapacitors *New Carbon Mater.* **36** 49–68
[8] [a] Lu Q, Chen J G and Xiao J Q 2013 Nanostructured electrodes for high-performance pseudocapacitors *Angew Chem. Int. Ed.* **52** 1882–9
[b] Bakker M G, Frazier R M, Burkett S, Bara J E, Chopra N, Spear S, Pan S and Xu C 2012 Perspectives on supercapacitors, pseudocapacitors and batteries *Nanomater. Energy* **1** 136–58
[9] [a] Bryan A M, Santino L M, Lu Y, Acharya S and D'Arcy J M 2016 Conducting polymers for pseudocapacitive energy storage *Chem. Mater.* **28** 5989–98
[b] Meng Q, Cai K, Chen Y and Chen L 2017 Research progress on conducting polymer-based supercapacitor electrode materials *Nano Energy* **36** 268–85
[c] Zha D, Xiong P and Wang X 2015 Strongly coupled manganese ferrite/carbon black/ polyaniline hybrid for low-cost supercapacitors with high rate capability *Electrochim. Acta* **185** 218–28
[10] [a] Yi C Q, Zou J P, Yang H Z and Xian L E N G 2018 Recent advances in pseudocapacitor electrode materials: transition metal oxides and nitrides *Trans. Nonferrous Met. Soc. China* **28** 1980–2001
[b] Frackowiak E 2013 Electrode materials with pseudocapacitive properties *Supercapacitors: Materials, Systems, and Applications* (Wiley) pp 207–37
[c] Yun X, Wu S, Li J, Li L, Zhou J, Lu P, Tang H and Zhu Y 2019 Facile synthesis of crystalline RuSe 2 nanoparticles as a novel pseudocapacitive electrode material for supercapacitors *Chem. Commun.* **55** 12320–3
[11] [a] Wu Z S, Zhou G, Yin L C, Ren W, Li F and Cheng H M 2012 Graphene/metal oxide composite electrode materials for energy storage *Nano Energy* **1** 107–31
[b] Jung J, Jeong J R, Lee J, Lee S H, Kim S Y, Kim M J, Nah J and Lee M H 2020 *In situ* formation of graphene/metal oxide composites for high-energy microsupercapacitors *NPG Asia Mater.* **12** 1–9
[12] [a] Chatterjee D P and Nandi A K 2021 A review on the recent advances in hybrid supercapacitors *J. Mater. Chem.* A **9** 15880–918
[b] Vlad A, Singh N, Rolland J, Melinte S, Ajayan P M and Gohy J F 2014 Hybrid supercapacitor-battery materials for fast electrochemical charge storage *Sci. Rep.* **4** 1–7

[13] [a] Wang Y, Wu X, Han Y and Li T 2021 Flexible supercapacitor: overview and outlooks *J. Energy Storage* **42** 103053
[b] Choi C, Lee J A, Choi A Y, Kim Y T, Lepró X, Lima M D, Baughman R H and Kim S J 2014 Flexible supercapacitor made of carbon nanotube yarn with internal pores *Adv. Mater.* **26** 2059–65
[c] Shao Y, El-Kady M F, Wang L J, Zhang Q, Li Y, Wang H, Mousavi M F and Kaner R B 2015 Graphene-based materials for flexible supercapacitors *Chem. Soc. Rev.* **44** 3639–65
[d] Palchoudhury S, Ramasamy K, Gupta R K and Gupta A 2019 Flexible supercapacitors: a materials perspective *Front. Mater.* **5** 83
[14] Dubal D P, Kim J G, Kim Y, Holze R, Lokhande C D and Kim W B 2014 Supercapacitors based on flexible substrates: an overview *Energy Technol.* **2** 325–41
[15] Reit R, Nguyen J and Ready W J 2013 Growth time performance dependence of vertically aligned carbon nanotube supercapacitors grown on aluminum substrates *Electrochim. Acta* **91** 96–100
[16] [a] Niu Z, Liu L, Zhang L, Zhou W, Chen X and Xie S 2015 Programmable nanocarbon-based architectures for flexible supercapacitors *Adv. Energy Mater.* **5** 1500677
[b] Chen T and Dai L 2014 Flexible supercapacitors based on carbon nanomaterials *J. Mater. Chem.* A **2** 10756–75
[17] Zhang Y Z, Wang Y, Cheng T, Lai W Y, Pang H and Huang W 2015 Flexible supercapacitors based on paper substrates: a new paradigm for low-cost energy storage *Chem. Soc. Rev.* **44** 5181–99
[18] Zhang H, Qiao Y and Lu Z 2016 Fully printed ultraflexible supercapacitor supported by a single-textile substrate *ACS Appl. Mater. Interfaces* **8** 32317–23
[19] [a] Kumar S, Saeed G, Zhu L, Hui K N, Kim N H and Lee J H 2021 0D to 3D carbon-based networks combined with pseudocapacitive electrode material for high energy density supercapacitor: a review *Chem. Eng. J.* **403** 126352
[b] Delbari S A *et al* 2021 Transition metal oxide-based electrode materials for flexible supercapacitors: a review *J. Alloys Compd.* **857** 158281
[20] [a] Mateen A, Javed M S, Khan S, Saleem A, Majeed M K, Khan A J, Tahir M F, Ahmad M A, Assiri M A and Peng K Q 2022 Metal-organic framework-derived walnut-like hierarchical Co-O-nanosheets as an advanced binder-free electrode material for flexible supercapacitor *J. Energy Storage* **49** 104150
[b] Li Q, Yue L, Li L, Liu H, Yao W, Wu N, Zhang L, Guo H and Yang W 2019 Metal-organic frameworks derived N, S co-doped bimetal nanocomposites as high-performance electrodes materials for supercapacitor *J. Alloys Compd.* **810** 151961
[21] Gupta A, Sardana S, Dalal J, Lather S, Maan A S, Tripathi R, Punia R, Singh K and Ohlan A 2020 Nanostructured polyaniline/graphene/Fe_2O_3 composites hydrogel as a high-performance flexible supercapacitor electrode material *ACS Appl. Energy Mater.* **3** 6434–46
[22] Wang Q, Ma Y, Liang X, Zhang D and Miao M 2019 Flexible supercapacitors based on carbon nanotube-MnO2 nanocomposite film electrode *Chem. Eng. J.* **371** 145–53
[23] Shown I, Ganguly A, Chen L C and Chen K H 2015 Conducting polymer-based flexible supercapacitor *Energy Sci. Eng.* **3** 2–26
[24] [a] Moon W G, Kim G P, Lee M, Song H D and Yi J 2015 A biodegradable gel electrolyte for use in high-performance flexible supercapacitors *ACS Appl. Mater. Interfaces* **7** 3503–11
[b] Li X, Lou D, Wang H, Sun X, Li J and Liu Y N 2020 Flexible supercapacitor based on organohydrogel electrolyte with long-term anti-freezing and anti-drying property *Adv. Funct. Mater.* **30** 2007291

[25] Presser V, Dennison C R, Campos J, Knehr K W, Kumbur E C and Gogotsi Y 2012 The electrochemical flow capacitor: A new concept for rapid energy storage and recovery *Adv. Energy Mater.* **2** 895–902

[26] Campos J W, Beidaghi M, Hatzell K B, Dennison C R, Musci B, Presser V, Kumbur E C and Gogotsi Y 2013 Investigation of carbon materials for use as a flowable electrode in electrochemical flow capacitors *Electrochim. Acta* **98** 123–30

[27] Boota M, Hatzell K B, Beidaghi M, Dennison C R, Kumbur E C and Gogotsi Y 2014 Activated carbon spheres as a flowable electrode in electrochemical flow capacitors *J. Electrochem. Soc.* **161** A1078

[28] Sasi S, Murali A, Nair S V, Nair A S and Subramanian K R V 2015 The effect of graphene on the performance of an electrochemical flow capacitor *J. Mater. Chem.* A **3** 2717–25

[29] Hatzell K B, Beidaghi M, Campos J W, Dennison C R, Kumbur E C and Gogotsi Y 2013 A high performance pseudocapacitive suspension electrode for the electrochemical flow capacitor *Electrochim. Acta* **111** 888–97

[30] Dennison C R, Beidaghi M, Hatzell K B, Campos J W, Gogotsi Y and Kumbur E C 2014 Effects of flow cell design on charge percolation and storage in the carbon slurry electrodes of electrochemical flow capacitors *J. Power Sources* **247** 489–96

[31] Pal B, Yang S, Ramesh S, Thangadurai V and Jose R 2019 Electrolyte selection for supercapacitive devices: a critical review *Nanoscale Adv.* **1** 3807

[32] Jung H *et al* 2012 Transparent, flexible supercapacitors from nano-engineered carbon films *Sci. Rep.* **2** 773

[33] Say M G *et al* 2020 Spray-coated paper supercapacitors *npj Flex Electron.* **4** 14

IOP Publishing

Nanomaterials for Advanced Energy and Power Storage Devices

K R V Subramanian, N Sriraam, T Nageswara Rao and Aravinda C L Rao

Chapter 2

Emerging technological advancements in thermal energy storage

H S Balasubramanya, Niranjan Murthy and K R V Subramanian

Thermal energy storage (TES) is crucial for enabling low-carbon heating solutions to address the demand–supply divergence in renewable energy generation. Despite its potential, domestic TES adoption remains limited, primarily, to a tank of hot water. Existing studies and reviews often place emphasis on comparing loading materials, overlooking system-level performance and emerging technological advancements in the tested performance of TES from initial-level to final-level analysis. In this chapter the impact of novel heat storage methods are discussed and evaluated through simulations of various TES materials and integration options, significant reductions in densities of energy and increases in explicit costs are observed when comparing system-level analysis to material-level analysis. Direct heating shows better integration potential with TES due to its high operating temperatures, unlike solar thermal systems or heat pumps, they are restricted to lower temperatures. TES properties are replicated within a home heating economic framework, revealing limited economic potential for TES in heat pump applications, even with high energy densities. For heat pumps, low principal cost is prioritized, though present high cost rates due to the energy crunch to improve TES's economic feasibility. Conversely greater energy density is the most valuable TES parameter for direct heating, as it facilitates substantial demand shifting to low-tariff periods, especially in lower demand homes where high electricity requirements for heating are minimal.

2.1 Introduction

Energy storage devices 'charge' by absorbing energy from renewable generation sources or from the electricity grid. They 'discharge' when delivering the deposited energy back into the grid. This process typically involves power conversion devices

doi:10.1088/978-0-7503-4901-7ch2

converting electrical energy (DC or AC) into various forms, such as thermal, electrochemical, chemical, electrical, and mechanical.

Energy storage plays a vital role in storing excess energy from current renewable sources, like wind power and solar PV, until it is needed. This capability enhances the incorporation of non-conventional energy into the overall energy system, mitigating the mismatch between energy demand and supply [1].

Various energy storage systems, whether centralized or decentralized, employ different technological approaches. The European Association for Storage of Energy (EASE) categorizes these technologies into five classes: chemical, electrochemical, electrical, mechanical, and thermal.

There are three diverse thermal energy storage (TES) principles: latent heat storage, sensible heat storage, and thermochemical heat storage. These technologies store energy across an extensive range of temperatures, provide diverse temporal ranges, and meet the inconsistency of energy system requirements.

2.2 TES technologies

TES technologies include, latent heat storage, sensible heat storage and thermochemical heat storage. Each has distinct characteristics and applications:

a. **Thermochemical Heat Storage:**
   - **Principle**: Stores energy through chemical reactions and chemisorption.
   - **Advantages**: Highest heat storage density, three times that of latent heat storage and ten times that of sensible heat storage, can be stored for long period without heat loss, high quality thermal energy, non-toxic, safe, economical, no supercooling or phase separation problems. Currently in early development and not yet fully implemented.

b. **Thermal Energy Storage Application for Waste Heat Recovery (WHR):**
   - **Potential for Heat Recovery:** Industrial processes have great possibities for use in waste heat retrieval, as the majority of industrial waste heat remains unused and is released directly into the environment. Energy-intensive industries namely steel, cement, glass, food processing oil, and gas are particularly noteworthy due to their high energy utilization and high waste heat emissions.
   - **Technical and Economic Challenges:** The main obstacle to effective waste heat recovery is technical and economic complexity. WHR technologies are essential to improve energy efficiency and reduce environmental impact.
   - **Recovery Efficiency:** By combining latent heat storage systems (LHTES) with waste heat release systems, precise and exact estimation of waste heat recovery can be achieved. Reliant on the characteristics of the material for heat storage, the size of the processing industry, and environmental conditions, the quantity of waste heat recovered can range from 44% to 85%.

c. **Latent Heat Storage (LHTES) Systems:**
   - **Combining with Waste Heat Systems:** Combining LHTES with waste heat systems can significantly improve waste heat recovery.
   - **Recovery Efficiency:** Achievable recovery rates range from 44% to 85%, depending on the properties of the material of the thermal energy storage device, the extent of the industry, and environmental conditions. By addressing technical and economic challenges and deploying advanced TES technologies, industries can significantly improve their energy efficiency and reduce their environmental footprint.

d. **Sensible Heat Storage:**
   - **Principle:** This method comprises storing thermal energy by rising the temperature of a liquid or solid. The quantity of energy stored is proportional to the heat capacity of the storage medium and the change in temperature.
   - **Advantages:** Pollution-free, low-cost, and mature with large-scale implementations.
   - **Disadvantages:** Great heat loss and less storage density of heat.
   - **Common Materials:** Often uses constituents like, molten salts, water, or rocks.

e. **Latent Heat Storage:**
   - **Principle:** Latent heat storage systems store energy during phase transitions, such as melting and solidifying, at a constant temperature.
   - **Advantages:** These systems can store huge quantities of energy in a small volume and are particularly effective for applications requiring temperature stability.
   - **Common Materials:** Phase change materials (PCMs) like paraffin, salt hydrates, and metallic alloys are commonly used.

Applications and Challenges: TES technologies enable the successful incorporation of non-conventional energy into the energy system by storing excess energy from recurrent sources, namely solar and wind power, until it is needed. This reduces the reliance on fossil fuels and helps stabilize energy supplies. However, practical applications of TES pose technical challenges, especially for long-term storage. Long-term storage requires large capacities and carries a high risk of heat loss, so the materials selected must be reliable, affordable and environmentally friendly. In spite of these challenges, current technological and research advances remain to progress the efficiency and profitability of TES systems, promising a more sustainable energy future.

Larger storage systems are needed to accommodate vast amounts of energy due to the low energy storage density of sensible thermal energy storage materials like brick, rock, concrete, and soil, which limits their practical applications. In contrast, PCMs undergo phase transitions, offering significant advantages. PCMs have a high energy storage capacity, absorbing, storing, and releasing substantial amounts of latent heat within a specific temperature range, as shown in figure 2.1. Even after undergoing thousands of phase change cycles, PCMs can retain latent heat energy

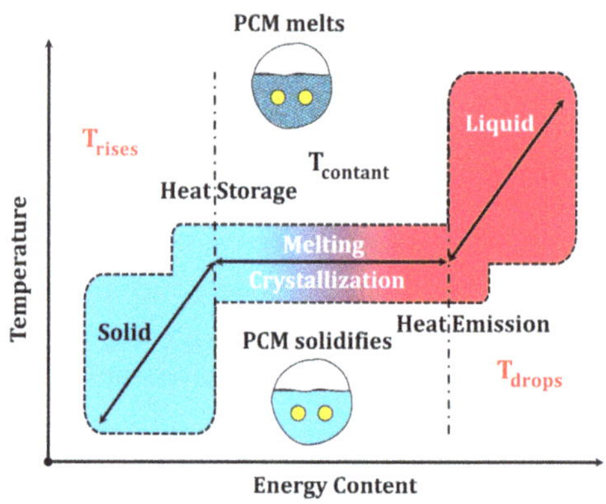

Figure 2.1. Phase transition diagram of PCM. Reprinted from [2], copyright (2024), with permission from Elsevier.

without degradation. The sensible and latent heat stored in PCMs can be calculated using equations (2.1) and (2.2) [2]:

1. Sensible heat Q_s is given by:

$$Q_s = m \cdot C_p \cdot \Delta T \tag{2.1}$$

   where:
   Q_s: Sensible heat
   m: Mass
   C_p: Specific heat capacity
   ΔT: Temperature change

2. Latent heat Q_l is expressed as:

$$Q_l = m \cdot h_{fg} \tag{2.2}$$

   where:
   Q_l: Latent heat
   m: Mass
   h_{fg}: Latent heat of fusion

In general, the enthalpy change of PCM and phase change temperatures during transition are measured using a differential scanning calorimeter (DSC). During DSC testing, a controlled dynamic heating power is applied to the sample, increasing its temperature at a constant rate. The resulting heating power is recorded and plotted to produce DSC curves. To accurately measure the 'net' vigorous heating power of the sample, material with a persistent specific heat capacity, such as indium metal or alumina is used.

The latent heat thermal energy storage method is particularly effective for air conditioning and solar heating applications because it can store a significant amount of energy at a constant temperature during phase transitions. Latent heat refers to the energy released or absorbed when a material changes phase from liquid to solid or vice versa. The phase change temperature can be optimized by selecting a PCM with a phase change temperature that enhances the temperature gradient between the material and the surrounding environment. For example, with paraffins and alkanes, the phase change temperature can be adjusted by altering the number of carbon atoms or creating diverse molecular alloys, allowing for a nearly constant range of phase change temperatures.

PCMs are essential for effective latent heat storage, offering several key properties for optimal performance in thermal energy storage systems [3]:

- **High Latent Heat of Fusion per Unit Mass**: A high latent heat of fusion allows a smaller amount of PCM to store a substantial amount of energy.
- **High Specific Heat**: This ensures additional sensible heat storage and helps minimize subcooling.
- **High Thermal Conductivity**: This facilitates efficient heat transfer by requiring a smaller temperature gradient to charge the storage material.
- **High Density**: High-density PCMs allow for a more compact storage solution, optimizing space utilization.
- **Appropriate Melting Point**: The PCM should have a melting point that aligns with the desired operating temperature range for the application.
- **Safety**: The PCM must be non-flammable, non-explosive, and non-poisonous to ensure safe handling and operation.
- **Chemical Stability**: The PCM should remain chemically stable to guarantee a long lifespan for the storage system.
- **Non-corrosiveness**: The PCM should not be corrosive to construction materials, preventing degradation of storage containers and system components.
- **Minimal Supercooling**: The PCM should exhibit minimal supercooling during freezing to maintain consistent performance and reliability.

Thermochemical Heat Storage: Utilizes reversible chemical reactions to store and discharge thermal energy. When energy is stored, a chemical compound absorbs heat and dissociates into its components. Upon energy demand, these components recombine, releasing heat. Thermochemical storage is known for its high energy density and potential for long-term energy storage with minimal losses.

Thermal Energy Storage (TES): addresses the irregular nature of solar energy. Developed extensively during the energy crisis of the 1970s, TES allows for the storage of solar energy to meet heating and hot water needs year-round. TES systems are categorized into long-term and short-term storage, with the latter including seasonal thermal energy storage (STES). STES, studied since the 1960s, stores excess summer heat to offset winter heating needs, reducing energy costs, fossil fuel use, and environmental impact [4].

Energy Storage Challenges in the Transition: Thermal energy storage includes a range of technologies capable of storing energy at temperatures from −40 °C to over 400 °C, including latent heat, sensible heat, and chemical energy. Sensible heat storage relies on the specific heat and heat capacity of the medium, often stored in well-insulated tanks. Water is commonly used in industrial and residential applications. Underground storage for large-scale applications also utilizes sensible heat in liquid and solid media but is limited by the medium's specific heat capacity.

PCMs offer higher storage capacity due to latent heat, maintaining a constant discharge temperature during phase changes. Thermochemical storage can achieve even higher capacities, operating at temperatures above 300 °C and offering efficiencies from 75% to nearly 100%. While sensible heat storage systems are commercially available, PCM and thermochemical storage systems are primarily used in the development and demonstration phase [5].

2.3 Performance and costs

- Sensible heat storage systems: Capacities of 10–50 kWh t^{-1} with efficiencies of 50%–90%, the amount ranges from 0.1 to 10 € kWh^{-1}.
- PCM systems: Offers higher capacities and efficiencies (75%–90%), with costs ranging from 10 to 50 € kWh^{-1}.
- Thermochemical storage systems: Can achieve capacities up to 250 kWh t^{-1} and efficiencies of 75% to nearly 100%, with costs estimated at 8–100 € kWh^{-1}.

The widespread adoption of TES technologies can significantly reduce CO_2 emissions and reliance on fossil fuels. In Europe, it could save around 14 lakh GWh per year and avoid 400 million tons of CO_2 emissions in the building and industrial sectors.

2.4 Barriers to adoption

1. **High Implementation Costs**: Despite decreasing battery costs, high initial investments remain a barrier. Combining TES with efficient district cooling and heating systems can reduce costs and improve system efficiency.
2. **Lack of Standardization**: The absence of uniform standards complicates long-term projects. ARANER addresses this by ensuring scalable and easily expandable systems.
3. **Outdated Regulatory Policies**: Regulatory frameworks lag behind technological advancements, particularly for residential energy storage. Comprehensive regulations are needed to support widespread adoption.

Addressing these challenges is crucial for advancing the energy transition and achieving greater energy efficiency.

2.5 Applications of sensible thermal energy storage

Hot water tanks are a well-established technology for storing thermal energy, offering significant energy savings in solar hot water heating and combined heat and power systems. These tanks are cost-effective and their efficiency is enhanced by optimal water stratification and highly effective insulation. Ongoing research and development efforts are focusing on innovations such as vacuum super insulation,

Table 2.1. Typical parameters of thermal energy storage systems a case study: 'Am Ackermannbogen' solar district heating system in Munich, Germany [11].

TEM system	Capacity ($kWh\ t^{-1}$)	Power (MW)	Efficiency (%)	Storage period (h, d, m)	Cost (€ kWh^{-1})
Sensible (hot water)	10–50	0.001–10	50–90	d/m	0.01–10
PCM	50–150	0.001–1	75–90	h/m	10–50
Chemical reactions	120–250	0.01–1	75–100	h/d	8–100

which achieves a heat loss rate of $\lambda = 0.01\ \ W\ \ mK^{-1}$ at 90 °C and 0.1 mbar, as well as improved system integration [6].

Hot water tanks, serving as buffer stores for hot water preparation, typically range from 500 l to several cubic meters in volume. They are also used in solar thermal systems combined with building heating systems (combined solar systems). In larger applications, such as seasonal storage for solar heat, these tanks can reach volumes of several thousands of cubic meters. The charging temperature for these systems is generally between 80 °C and 90 °C. By employing a heat pump for discharge, the usable temperature difference can be increased to around 10 °C.

An example of this technology in use is the solar district heating system 'Am Ackermannbogen' in Munich, Germany. This system offers solar energy for space heating and domestic hot water to about 320 apartments in 12 multi-story buildings, covering around 30 400 m^2 of living space. It is designed to meet over 50% of the annual heat demand (approximately 2000 MWh per year) using solar energy collected by 2761 m^2 of flat-plate collectors. The heat collected is either used directly or stored in a 6000 m^3 underground seasonal hot water storage tank. Additional heating is supplied by an absorption heat pump powered by the city district heating system, which uses the seasonal storage as a low-temperature heat source. This configuration allows for a wide range of storage operation temperatures, from 10 °C to 90 °C. The parameter values for TES system are indicated in table 2.1 [11].

Connecting the district heating system directly to the building heating installations helps to avoid typical temperature drops at heat exchangers and raises the temperature spread. The district system operates with a source temperature of 60 °C and a return temperature of 30 °C, both of which are carefully observed. In the second year of operation, the solar energy fraction reached 45%, with the potential to exceed 50% through further optimization.

Latent and sensible thermal energy storage solutions

Thermal energy storage (TES) plays a vital role in balancing energy supply and demand, particularly in systems that utilize renewable energy sources. TES solutions can be broadly characterized into latent heat storage and sensible heat storage. Each method has its own characteristics, applications, advantages, and disadvantages [7]. The following is a comparison of these two thermal energy storage approaches:

Sensible thermal energy storage (STES)

Principles and Technology Sensible thermal energy storage comprises storing thermal energy by rising the temperature of a storage medium without changing its phase. Common storage media include water, rock, and molten salt. The stored heat is retrieved by cooling the medium.

Characteristics of STES [8]

- **Storage Medium**: Typically uses water, rock, or molten salt. Energy can be stored at temperatures ranging from −40 °C to over 400 °C.
- **Energy Density**: Lower compared to latent heat storage, with a capacity of 10–50 kWh t^{-1}. Storage efficiency ranges from 50% to 90%, depending on the medium and insulation.
- **Applications**: Commonly used in residential hot water systems and for large-scale seasonal storage in district heating systems.
- **Solar Thermal Systems**: Employed in solar thermal systems for heating and hot water.
- **Simplicity and Reliability**: Well-established technology that is relatively easy to implement.
- **Cost-Effectiveness**: Generally lower initial costs compared to latent heat storage.
- **Scalability**: Suitable for both small and large applications.
- **Limitations**: Requires larger storage volumes to store equivalent amounts of energy. Efficiency depends on the specific heat and insulating properties of the storage medium.

Latent heat energy storage (LTES)

Principles and Technology Latent heat storage involves storing thermal energy by changing the phase of a storage medium, known as a phase change material (PCM). Energy is absorbed or released during the phase transition.

Characteristics of LTES

- **Storage Medium**: Utilizes phase change materials (PCMs) such as paraffin, salt, metal alloys, etc.
- **Temperature Range**: Designed for specific phase transition temperatures.
- **Energy Density**: Higher than sensible heat storage, with capacities of 50–150 kWh t^{-1}. Efficiency ranges from 75% to 90%, thanks to the high latent heat associated with phase changes.
- **Applications**: Used in building HVAC systems, refrigerated warehouses, and integrated with solar power systems to manage and store heat.
- **High Energy Density**: Stores more energy per unit volume compared to sensible heat storage [9].

- **Stable Discharge Temperature**: Maintains a constant temperature during the phase change, beneficial for applications requiring precise temperature control.
- **High Cost**: Generally higher initial costs due to the expense of PCM and related technology.
- **Material Stability**: Long-term stability of PCMs can impact performance and reliability [10].
- **Complexity**: Requires careful control of phase transitions and heat transfer, making the system more complex.

Summary

Sensible Heat Storage (STES) is known for its simplicity, cost-effectiveness, and scalability but has lower energy density and requires larger storage volumes. It is suitable for an extensive range of applications, including residential and large-scale systems.

Latent Heat Storage (LTES) offers higher energy density and stable discharge temperatures, making it perfect for applications necessitating exact temperature control. However, it tends to be more complex and expensive due to the cost of PCMs and the need for careful system management.

Both storage methods have their individual set of advantages and are selected based on the definite requirements of the application, including energy density needs, temperature stability, cost considerations, and system complexity. Table 2.2 provides the comparative summary of the two technologies [11].

Both sensitive and latent thermal storage solutions have their own strengths and are suitable for different applications. Sensitive thermal storage is cost-effective and reliable for large scale, simple applications, while latent thermal storage offers higher energy density and precise temperature control, making it ideal for specialized or

Table 2.2. Comparative summary [11].

Feature	Sensible thermal energy storage	Latent thermal energy storage
Storage medium	Water, rocks, molten salts	Phase change materials (paraffin, salts)
Energy density	10–15 kWh t^{-1}	50–150 kWh t^{-1}
Efficiency	50%–90%	75%–90%
Temperature range	−40 °C–400 °C	Specific to phase change temperatures
Cost	€0.1–€10 per kWh	€10–€50 per kWh
Applications	Domestic hot water, district heating, solar thermal systems	Building HVAC, cold storage, solar energy systems
Advantages	Simplicity, reliability, cost-effectiveness, scalability	High energy density, stable discharge temperature
Disadvantages	Lower energy density, temperature limitations	High costs, material stability issues, complexity

space-constrained applications. The choice between these TES solutions are subject to the explicit requirements of the application, including cost, energy density, efficiency, and temperature stability.

References

[1] *LDES Council: Net-zero Heat: Long Duration Energy Storage to Accelerate Energy System Decarbonization* (McKinsey & Company, 2022) https://mckinsey.com/capabilities/sustainability/our-insights/net-zero-heat-long-duration-energy-storage-to-accelerate-energy-system-decarbonization

[2] Paul U K, Mohtasim M S, Kibria M G and Das B K 2024 Nano-material based composite phase change materials and nanofluid for solar thermal energy storage applications: Featuring numerical and experimental approaches *J. Energy Storage* **98** 113032

[3] UK: flexibility for low carbon electric heating the role of smart thermal storage in providing flexible, low carbon electric heating 2022 https://static1.squarespace.com/static/61af1582256 18e1511f3323f/t/6348080d3e4c696e22f6b8ca/1665665039164/Benefits

[4] EU 2019 Directive 2019/944 of the EU parliament and the council on common rules for the internal market for electricity. Article 2, Paragraph (59) https://eur-lex.europa.eu/legal-content/EN/TXT/ ? uri = celex%3A32019L0944

[5] EASE: Energy Storage Technologies 2023 https://ease-storage.eu/energy-storage/technologies/

[6] Sample System Design and Interconnection, The Pew Charitable Trusts: Energy Smart Technologies in the Evolving Power System 2016 https://pewtrusts.org/-/media/assets/2016/02/the_smart_grid_how_energy_technology_is_evolving_print.pdf

[7] EASE 2021 Ancillary services—energy storage applications forms https://ease-storage.eu/wp-content/uploads/2021/08/Ancillary-Services.pdf

[8] *EnTEC: study on the performance of support for electricity from renewable sources granted by means of tendering procedures in the Union* 2022 https://op.europa.eu/en/publication-detail/-/publication/e04f3bb2–649f-11ed-92ed-01aa75ed71a1/language-en

[9] Directive (EU) *2018/2002 of the European Parliament and of the Council of 11 December 2018 amending Directive 2012/27/EU on energy efficiency* https://eur-lex.europa.eu/legal-content/EN/TXT/ ? uri = uriserv:OJ.L_.2018.328.01.0210.01.ENG

[10] Statista *Market size of paper and pulp industry worldwide from 2021 to 2029* https://statista.com/statistics/1073451/global-market-value-pulp-and-paper/

[11] Schmidt T, Mangold D, Muller-Steinhagen H 2003 Seasonal thermal energy storage in Germany *ISES Solar World Congress (Goteborg, Schweden)* pp 14–19

Chapter 3

Quantum dots as an efficient supercapacitor for the modern world

Janardan Sannapaneni, H Manjunatha, K Venkata Ratnam and B Ramakrishna Rao

Complex environmental problems such as global warming and fossil fuel depletion are gradually increasing and motivating researchers to search for new alternative materials for energy conversion and storage, where zero-dimensional quantum dots based on carbon (CQDs) and graphene (GQDs) occupy a significant position in storing the energy due to their properties such as chemical inertness, non-toxicity, low cost, and easy functionalization. In addition, abundant research works prominently describe the applications of quantum dots in batteries, supercapacitors, and composite materials. In this chapter, we highlight recent advances in the synthesis of quantum dots and their applications in various energy storage devices with respective mechanistic aspects.

3.1 Introduction

Renewable energy resources such as wind energy, geothermal energy, and solar energy, which are ecofriendly due to the negligible carbon emissions are currently attracting more attention for their role in helping with complex environmental issues such as global warming, climate change, and energy crises [1]. Moreover, storing this harvested energy is one of the greatest challenges [2] being solved by batteries and supercapacitors [3–6]. Quantum dot-based (QD) storage systems occupy a significant position due to their easy fabrication, specific surface area, non-toxicity, low cost, and easy functionalization [7–9]. Moreover, quantum dots are generally spherical to quasispheroidal and have sizes ranging from 0 to 10 nm [10, 11]. Synthetic types of QDs are broadly classified as top-down [12–14] and bottom-up approaches [15–17]. Zero-dimensional QDs are significant due to their high quantum confinement and that they obtain high bandgap and high surface to volume ratio with ultra-small size [18, 19]. Moreover, one can alter properties such

doi:10.1088/978-0-7503-4901-7ch3

as size, shape, functional groups of QDs with active sites and used as active electrodes [20, 21], current collectors [22, 23] with high conductivity [24], large capacity [25], and good cycle stability [26], with improved electrochemical energy. The efficient electron donor and acceptor properties of QDs make them the best candidates for hydrogen evolution, photocatalytic hydrogen evolution, and oxygen evolution reactions [27–29]. In addition, QDs doped with heteroatoms show good adsorption properties which is also promising for hydrogen storage [30].

In this chapter we particularly focus on the synthesis of carbon and graphene-based quantum dots and heteroatom doping and its applications in energy storage devices as supercapacitors with its specific capacitance with other reaction parameters.

3.2 Synthesis of carbon quantum dots

Carbon quantum dots (CQDs) are generally synthesized using two synthetic approaches, i.e. the bottom-up and top-down approaches as shown in figure 3.1. In the top-down approach one can break down the carbon macromolecules to microstructures such as fullerene, graphene, carbon black and carbon nanotubes under specific intense conditions through mechanical milling, chemical oxidation, electrochemical oxidation, ultrasonic, and the arc discharge method. In the bottom-up approach small macromolecules are converted into macromolecules through mild reaction approaches such as plasma treatment, microwave assisted treatment, hydrothermal method, pyrolysis, and ultrasonic reactions [31]. Moreover, the surface area and electrical conductivity properties of carbon dots, especially photoluminescence, depends upon the synthetic process where it will show higher PL in the bottom-up approach than top-down approach. One can enhance the PL of molecules by incorporating negatively charged polar functional moieties such as –OH, –COOH, $-NH_2$, –CO, and $-CONH_2$. In addition to functionalization, doping with heteroatoms also leads to improvement of quantum yield observed with the incorporation of heteroatoms like O, N, which makes this class of candidate best suited to supercapacitor applications.

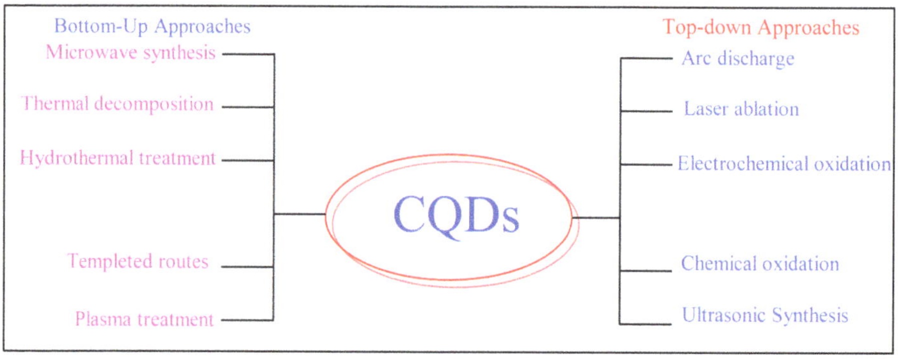

Figure 3.1. Synthetic approaches of CQDs.

Table 3.1. General synthetic approaches of CQDs from natural precursors [33].

Supercapacitors	C-dot-based material	Precursors used	Synthetic methodology	Specific function
1	C-dots/rGO	Citric acid	Hydrothermal	Charge storage
2	GQDs/AC	Coal	Oxidation	Charge storage
3	N-GQD/GH/CF	Pyrene	Molecular fusion	Charge storage
4	MnO_2/C-dots/GA	Milk	Hydrothermal	Charge storage
5	MnO_2/C-dots/GA	Sucrose	Oxidation	Charge storage
6	MnO_2/C-dots/GA	Milk	Hydrothermal	Charge storage
7	S-CQD/PANI	Sucrose	Oxidation	Charge storage
8	Ppy/C-dots	Ascorbic acid	Pyrolysis	Charge storage

In general, the products obtained from the top-down approach show a well-defined architecture and simple process and in the case of the bottom-up approach many byproducts are produced in addition to our target molecule due to the intense reaction conditions [32]. However, the bottom-up approach has its own significance in doping and surface passivation of the material without having any post synthetic treatments. Considering the above methodologies scientists are looking towards using natural sources to produce the CQDs as mentioned in the table 3.1.

3.3 Synthesis of graphene quantum dots

3.3.1 Top-down approach

In general, several methods are used in the top-down approach to synthesize GQDs such as electrochemical, hydrothermal, solvothermal, strong acid oxidation, stripping, and electron beam irradiation [34, 35]. However, among them the hydrothermal and solvothermal methods have more significance in producing high yields [36].

In general, GQDs were synthesized by degrading the carbon with base and ammonia. Graphene oxide also converted to biocompatible GQDs through a simple solvothermal method with the help of DMF and nitrogen source. The size and shape dependent GQDs were synthesized using the electrochemical method and high quality GQDs were prepared using chemical vapor deposition using 3D graphene [37]. In addition, multiwalled carbon nanotubes were also used for the synthesis of size controlled GQDs and they generally have narrow particle sizes and thicknesses greater than 1.25 nm.

Through electrochemical oxidation GQDs were prepared by Pillai *et al* using $LiClO_4$ dissolved propylene carbonate as an electrolyte [38]. GQDs with sizes less than 100 nm with the strong oxidation of C–C bonds and graphene sheets were created with the help of conc. H_2SO_4 and conc. HNO_3.

3.3.2 Bottom-up methods

This method mainly focuses on the synthesis of GQDs from the aggregation of simple molecules through microwave, ultrasonic, solvent chemistry methods, and thermal decomposition of polycyclic aromatic hydrocarbons. Nitrogen and sulfur doped GQDs were prepared in an aqueous system. Salicylic acid assisted GQDs were synthesized through free radical polymerization which releases simple by products such as H_2O and CO_2 [39].

Nitrogen–sulfur doped GQDs were prepared by mixing citric acid and thiourea or urea.

Pyrolytic citric acid was also used in the preparation of GQDs. Wang *et al* found that a sodium catalyzed reduction of methyl benzene and hexabromobenzene could produce hybrids of GQDs and carbon nanotubes.

In recent years a combination of both approaches (top-down and bottom-up) have attracted more attention. The graphitized carbon powder synthesis from calcining organic molecules is the bottom-up approach and using this carbon in the synthesis of GQDs is the top-down approach. Similarly, carbon powder is synthesized from neem leaves and it is utilized it for the synthesis of GQDs using sulfuric acid and nitric acid.

3.3.3 Heteroatom doping

The doping of material with heteroatoms results in significant change in the electronic structure of a material. The electronic structure of graphene can be changed by introducing heteroatoms which either donate or accept electrons. Doping of GQDs involves the replacement of C with the other heteroelements such as N, Se, S, P, B, Si, and I [40–44]. Moreover, the non-metallic heteroatoms especially N and B effectively replace the C in graphene and enhance the migration of electron holes, which makes GQDs an effective material in electrochemical capacitors. Moreover, N doping improves the conductivity in graphene, and is done through chemical vapor deposition, thermolysis, and plasma treatment [41–45].

Carbonaceous materials such as graphene, graphite, activated carbon, and carbon aerogel play a major role in the synthesis of SCs and electric double layer capacitors, asymmetric SCs, micro SCs, and hybrid SCs and their importance is summarized in table 3.2, which promotes the use of green energy using advanced materials. GQDs are shown to be excellent devices for energy storage as super-capacitors for their physical and chemical properties, size, shape will enhance the electrochemical performance [46–62].

3.4 Carbon quantum dots for supercapacitors

Due to lower energy density the practical applications of SCs are reduced, moreover one can increase the energy density by increasing OPW or Csp based on the following relation $E = ½C \times V_2$. However, the addition of transition metal oxides such as Fe_2O_4, MnO_2, RuO_2, and $NiCo_2O_4$ will improve the energy density but due to their low electrical conductivity carbonaceous materials are added to overcome the above problems. Reduced carbon dots are obtained by oxidative exfoliation of graphite material in an inert atmosphere. The reduced carbon dots with RuO_2 synthesized using the solgel method show significant electrochemical performance where Csp with a value of 460 F g^{-1} has been obtained with the current density of 50 A g^{-1}.

Table 3.2. Various synthetic methods used for the synthesis of GQDs [62].

Synthesis method	Precursor	Particle size (nm)	Device	Effect (capacitance)
Direct chemical cutting	GO	1–5.4	Micro-SC	100%
Direct chemical cutting	GO	1–6	Micro-SC	30%
Direct chemical cutting	Nanographite	3–6	ASC	20%
Electrochemical	Graphite	5	3-electrode	4 times
Electrochemical	Graphite	3–8	Symmetric SC	60%
Hydrothermal	Citric acid and thiourea	1–5	3-Electrode	—
Hydrothermal	Polyethylene glycol-400	4–7	EDLC	34%
Electrochemical	Graphite	2–5	EDLC	200%
Cage opening	Fullerene	7–8	EDLC	—
Hydrothermal	Citric acid and ethylene diamine	3–5	EDLC	74.3%
Hydrothermal	1,3,6-Trinitropyrene and hydrazine hydrate	4–5	Solid state flexible SC	—
Hummers	GO	2–4.5	ASC	17%
Direct chemical cutting	Coal powder	2.9	EDLC	67%
Hydrothermal	Citric acid	20–30	3-Electrode	33%
Direct chemical cutting	Graphite powder	3–7	3-Electrode	100%
Hydrothermal	Citric acid	3–10	3-Electrode	32%

The conductive nature of C-dots due to their enhanced ionic transport during discharge and charge process and oxygen incorporation on C-dots improves the use of RuO_2 improving the redox reactions.

C-dots showed significant and unique properties in the field of electrochemistry when compared to other materials due to the super capacitive activities of different materials listed in table 3.3, where C-dots have greater roles in SC applications [58, 61].

The detail of the CQD synthesis technique is presented in table 3.2. Here, only the most typical techniques are described and illustrated with literary samples. Due to space restrictions and their lesser importance to our subject than the methods described above, there are several additional methods that are not included here.

Due to their exceptional characteristics, such as their exceptional electron conductivities, high surface areas, changeable band gaps, and remarkable wettability in various solvents, electrodes based on CDs can offer ultrahigh capacities and maximal efficiencies. For instance, GQDs were used by Niu *et al* in place of conventional carbon black (CB) as the nanoscale conductive agents, which showed

Table 3.3. Electrochemical performance of carbon dots and their derivatives as SCs with specific parameters.

Material	Specific capacitance ($F\ g^{-1}$)		Current density ($A\ g^{-1}$)	Energy density ($Whkg^{-1}$)	Power density ($W\ kg^{-1}$)	References
	Unmodified	Modified with C-dots				
Co_3O_4	1446	1603	1	78.4	810	[63]
NiS	710	880	2	30	33 000	[64]
NiO/Co_3O_4	—	1775	1	21.3	16 000	[65]
Bi_2O_3	—	343 C g^{-1}	0.5	32	8400	[61]
Bi_2O_3	849	1046	1	79.9	770.9	[66]
$NiCo_2O_4$	699	2202	1	73.5	499.8	[67]
CoS/rGO	—	697	1	36.6	16 000	[68]
Polypyrrole	267	576	0.5	30.1	250	[69]
V_20_5	60	270	1	60	4100	[70]
PANI	182.5	222.7	1	—	—	[71]
$CuCo_2O_4$	464.4	779.8	1	39.5	1203.7	[58]
PVA-PEDOT	161.48	291.86	100 mVs^{-1}	16.95	—	[72]
MOS_2	100	323.5	1	38.47	489.96	[73]
$MnCo_2O_{4.5}$	368	1625	1	24	2800	[74]
$CuCo_2S_4$	908	1725	0.5	—	—	[70]
NiCoS	562	678	0.2	—	—	[75]
$Ni(OH)_2$	1414.3	2750	1	57.4	683.7	[76]

great promise for the creation of an efficient conductive network to increase the specific capacitances of SCs. Compared to carbon black, an electrode material with GQDs as the conducting agent produced better specific capacitance and improved rate performance. Additionally, CDs offer significant benefits for the development of flexible and micro-supercapacitors (MSCs). CDs can be made in such nanoscale sizes so that a variety of electrode forms can be created such as micro-supercapacitors and flexible (MSCs). It is possible to build electrodes with a variety of forms and shorten the ion transmission channels thanks to the nanoscale diameters of CDs. In comparison to porous carbon and graphene, CDs with large specific surface areas and functional groups can offer greater capacities. Due to their excellent stability, they can be used with a variety of electrode production techniques, including inkjet printing, suction filtering, and electroplating (EPD). Table 3.4 compiles information gathered from the comprehensive literature publications on the electrochemical performance of CDs and their derivatives for SCs.

Large specific surface area and nitrogen doping can help N-GQDs operate better as supercapacitors than regular carbon materials. For instance, Liu *et al*'s GQD/MoS_2-QD asymmetric MSC demonstrated a rapid frequency response with a modest relaxation time constant of 0.087 ms and a robust capacity retention of

Table 3.4. Electrochemical performance of graphene and their derivatives as efficient SCs.

Electrode material	Synthetic methodology	Specific capacity ($F g^{-1}$)	References
GQDs	Liquid phase oxidation of bituminous coal	230	[78]
GDQs	Oxidation of carbon fiber	213	[79]
GDQs	Commercially purchased	2220	[80]
GDQs	Pyrolysis of citric acid	363–216	[81]
GDQs	Nitric acid treatment on graphene oxide	296.7	[82]
GDQs	From carbon nanofibre	335	[83]
GDQs	Chemical oxidation of bituminous coal	315	[84]
Nitrogen doped GDQs	Commercially purchased	344	[85]
Iron and sulfur doped atoms with GDQs	Hydrothermal method	476.2	[86]
N-Doped GDQs/GH/CF	Molecular fusion strategy using pyrene	−93.7	[47]
N-Doped His-GQD/LDH	Microwave reactor	1526	[87]
N,F-GDQs	Refluxing system	270	[88]
PANI/GDQ-rGO	Carbon fiber exfoliation by mixed acids	1036	[89]
PVA-GDQ-Co_3O_4 composite with PEDOT	Commercially purchased	361.97	[90]
Ferrocenyl modified GDQs	Heating of citric acid at 200 °C	284.1	[91]
PAN-PANI@GDQs	Hydrothermal autoclave method from rice power	105–587	[92]
$NiCo_2O_4$@GDQs	Heating of GO at 200 °C for 24 h	1242	[93]
GQD/$CuCo_2S_4$	Commercially available	1725	[94]
GQDs-MoS_2	Pyrolysis of citric acid	380	[95]
$Ni(OH)_2$/af.GQDs	Exfoliated graphene	2653	[96]
Zucchini-derived CDQs/RGO	—	218	[93]
I-doped graphene/CQDs	—	374	[93]
CDQs/graphene hydrogel	—	264	[46]
N,S-codoped CDQs/RGO	—	141	[97]
RGO/CDQs	—	262	[98]
CQDs/RGO	—	212	[99]
N-doped CQDs/RGO	—	278	[100]
HPCNs	—	230	[101]
Activated GQDs	—	236	[101]
GDQs	—	534.7 $\mu F cm^{-2}$	[22]
GDQs/MnO_2	—	1107.4 $\mu F cm^{-2}$	[22]
CQDs/RuO_2	—	594	[102]

(*Continued*)

Table 3.4. (*Continued*)

Electrode material	Synthetic methodology	Specific capacity ($F\ g^{-1}$)	References
Reduced CQDs	—	128	[102]
GQDs/CNT	—	44 mF Cm^{-2}	[103]
CQDs/$NiCo_2O_4$	—	856	[104]
GQDs/PANI	—	667.5 $\mu F\ cm^{-2}$	[78]
GQDs/PANI	—	210 $\mu F\ cm^{-2}$	[78]
GQDs/3D graphene	—	268	[105]
3D CQD aerogel	—	294.7	[106]
CQDs/graphene microfibers	—	607 mF cm^{-2}	[107]
CQDs/graphene microfibers	—	215mF cm^{-2}	[107]
$NiCo_2S_4$/GQDs	—	678.22	[75]

Abbreviations: GQDs, graphene quantum dots; GH, graphene hydrogel; CF, carbon fibers; LDH, layered double hydroxides; PANI, polyaniline; PVA, polyvinyl alcohol; PEDOT, poly(3,4-ethylene dioxythiophene); PAN, polyacrylonitirile; Af, amino-functionalized.

89.2% after 10 000 cycles. The voltage and current were twice as high when two microdevices were run in series and parallel, proving that the capacitor series and parallel connections' guiding principles were adhered to. Additionally, even at 100 V S^{-1}, the CV curves maintained their roughly rectangular shape, demonstrating the high-rate capability.

3.5 Graphene quantum dots for supercapacitors

Supercapacitors/micro-supercapacitors (SCs/MSCs) are gaining more attention as energy storage devices because of their high-power density, small size, quick charging and discharging times, and extended lifespans. Because of their various charge storage techniques, SCs/MSCs are essentially separated into electrochemical double layer capacitors and pseudo capacitors. Typically, the mechanism of the electrochemical double layer type of SCs/MSCs with GQDs as the electrode material is charge adsorption on the electrode surface. The resulting MSC has a specific capacitance of 534.7 F cm^2, a rate performance of up to 1000 V s^{-1}, and a specific capacitance of 97.8% after 5000 cycles in comparison to its initial specific capacitance. This was achieved by electrodepositing GQDs on a gold finger electrode.

With a specific capacitance of 534.7 F cm^2, a rate performance of up to 1000 V s^{-1}, and a specific capacitance of 97.8% after 5000 cycles in comparison to its initial specific capacitance, the produced MSC offers impressive performance characteristics. In addition, the MSC also possesses a quick relaxation time constant ($0 = 103.6$ s in an aqueous electrolyte and $0 = 53.8$ s in an ionic liquid electrolyte). Using chelated graphene and GQDs, Lee *et al* created a flexible, transparent MSC with a specific capacity of 9.09 F cm^2, a high transmittance of 92.97% at a wavelength of 550 nm, and highly bendable properties [69]. It is nevertheless possible to maintain long cycle stability in the bend scenario [58, 70–72, 77].

3.6 Conclusion

In conclusion, this chapter mainly emphasizes the importance and great potential of quantum dot composites (CQDs and GQDs) in the development of high-performance energy storage systems especially supercapacitors. It is reasonable to conclude that QDs are becoming important multifunctional materials for energy storage/conversion devices. With the development of advanced technologies and characterization methods, new physicochemical properties of QDs will be discovered, which will lead to an extension of their application to other promising research fields. We can foresee that the research of advanced electrode materials based on QDs and their potential applications in energy-related fields will encounter an explosive growth soon.

References

[1] Zhao N and You F 2020 *Appl. Energy* **279** 115–889

[2] Czaun M, Kothandaraman J, Goeppert A, Yang B, Greenberg S, May R B, Olah G A and Prakash G K S 2016 *ACS Catal.* **6** 7475–84

[3] Yang Y, Bremner S, Menictas C and Kay M 2018 *Renew. Sustain. Energy Rev.* **91** 109–25

[4] Denholm P, Nunemaker J, Gagnon P and Cole W 2020 *Renew Energy* **151** 1269–77

[5] Ma T, Yang H and Lu L 2015 *Appl. Energy* **153** 56–62

[6] Dubal D P, Ayyad O, Ruiz V and Gómez-Romero P 2015 *Chem. Soc. Rev.* **44** 1777–90

[7] Ansari S *et al* 2022 *Nanomaterials* **12** 3814

[8] Fowley C, Nomikou N, McHale A P, McCaughan B and Callan J F 2013 *Chem. Commun.* **49** 8934–6

[9] Wu P, Xu Y, Zhan J, Li Y, Xue H and Pang H 2018 *Small* **14** 180–1479

[10] Wang J, Tang L, Zeng G, Deng Y, Dong H, Liu Y, Wang L, Peng B, Zhang C and Chen F 2018 *Appl. Catal. B:* **222** 115–23

[11] Molaei M J 2019 *RSC Adv.* **9** 6460–81

[12] Ahmad P *et al* 2019 *Ceram. Int.* **45** 22765–8

[13] Yang J, Ling T, Wu W T, Liu H, Gao M R, Ling C, Li L and Du X W 2013 *Nat. Commun.* **4** 1–6

[14] Sun H, Ji H, Ju E, Guan Y, Ren J and Qu X 2015 *Chem. Eur. J.* **21** 3791–7

[15] Yan Y, Zhai D, Liu Y, Gong J, Chen J, Zan P, Zeng Z, Li S, Huang W and Chen P 2020 *ACS Nano* **14** 1185–95

[16] Dervishi E, Ji Z, Htoon H, Sykora M and Doorn S K 2019 *Nanoscale* **11** 16571–81

[17] Gao T, Wang X, Zhao J, Jiang P, Jiang F-L and Liu Y 2020 *ACS Appl. Mater. Interfaces* **12** 22002–11

[18] Baker S N and Baker G A 2010 *Angew. Chem. Int. Ed.* **49** 6726–44

[19] Xu N *et al* 2020 *Adv. Sci.* **7** 2002–209

[20] Li J, Yun X, Hu Z, Xi L, Li N, Tang H, Lu P and Zhu Y 2019 *J. Mater. Chem.* A **7** 26311–25

[21] Li Z, Bu F, Wei J, Yao W, Wang L, Chen Z, Pan D and Wu M 2018 *Nanoscale* **10** 22871–83

[22] Zhao Z and Xie Y 2017 *J. Power Sources* **337** 54–64

[23] Liu W W, Feng Y Q, Yan X B, Chen J T and Xue Q J 2013 *Adv. Funct. Mater.* **23** 4111–22

[24] Tan H, Cho H W and Wu J J 2018 *J. Power Sources* **388** 11–8

[25] Huang S, Wang M, Jia P, Wang B, Zhang J and Zhao Y 2019 *Energy Storage Mater.* **20** 225–33
[26] Li X, Hu K, Tang R, Zhao K and Ding Y 2016 *RSC Adv.* **6** 71319–27
[27] Shi R, Li Z, Yu H, Shang L, Zhou C, Waterhouse G I N, Wu L Z and Zhang T 2017 *Chem. Sus. Chem.* **10** 4650–6
[28] Cao Y, Zhou H, Qian R C, Liu J, Ying Y L and Long Y T 2017 *Chem. Commun.* **53** 5729–32
[29] Liu J, Liu Y, Liu N, Han Y, Zhang X, Huang H, Lifshitz Y, Lee S T, Zhong J and Kang Z 2015 *Science* **347** 970–4
[30] Malcek M and Bucinsky L 2020 *Theor. Chem. Acc.* **139** 167
[31] Xia C, Zhu S, Feng T, Yang M and Yang B 2019 *Adv. Sci.* **6** 1901316
[32] Sciortino A, Cannizzo A and Messina F 2018 *C* **4** 67
[33] Mahmudul Hasan A M, Akib Hasan M, Reza M, Atek, Islam M and Abu Bin Hasan Susan M 2021 *Mater. Today Commun.* **29** 102732
[34] Malik R, Lata S, Soni U, Rani P and Malik R S 2020 *Electrochim. Acta* **364** 137281
[35] Zhu S, Meng Q, Wang L, Zhang J, Song Y, Jin H, Zhang K, Sun H, Wang H and Yang B 2013 *Angew. Chem. Int. Ed.* **125** 4045–9
[36] Xu Y, Jia X H, Yin , He X B and Zhang Y K 2014 *Anal. Chem.* **86** 12122–9
[37] Liu Q, Guo B, Rao Z, Zhang B and Gong J R 2013 *Nano Lett.* **13** 2436–41
[38] Ananthanarayanan A, Wang X, Routh P, Sana B, Lim S, Kim D H, Lim K H, Li J and Chen P 2014 *Adv. Funct. Mater.* **24** 3021–6
[39] Jeon S J, Kang T W, Ju J M, Kim M J, Park J H, Raza F, Han J, Lee H R and Kim J H 2016 *Adv. Funct. Mater.* **26** 8211–9
[40] Zhu J, Tang Y *et al* 2017 *ACS Appl. Mater. Interfaces* **9** 14470–7
[41] Qu D, Zheng M, Du P, Zhou Y, Zhang L, Li D, Tan H, Zhao Z, Xie Z and Sun Z 2013 *Nanoscale* **5** 12272–7
[42] Dong Y, Shao J, Chen C, Li H, Wang R, Chi Y, Lin X and Chen G 2012 *Carbon* **50** 4738–43
[43] Jin H, Huang H, He Y, Feng X, Wang S, Dai L and Wang J 2015 *J. Am. Chem. Soc.* **137** 7588–91
[44] Suryawanshi A, Biswal B, Mhamane D, Gokhale R, Patil S, Guin D and Ogale S 2014 *Nanoscale* **6** 11664–70
[45] Wang Z, Yu J *et al* 2016 *ACS Appl. Mater. Interfaces* **8** 1434–9
[46] Yang S, Sun J, He P, Deng X, Wang Z, Hu C, Ding G and Xie X 2015 *Chem. Mater.* **27** 2004–11
[47] Zhang Y, Zhao J, Sun H, Zhu Z, Zhang J and Liu Q 2018 *Sensors Actuators* B **266** 364–74
[48] Wei D, Liu Y, Wang Y, Zhang H, Huang L and Yu G 2009 *Nano Lett.* **9** 1752–8
[49] Han C P, Chen C J, Hsu C C, Jena A, Chang H, Yeh N C, Hu S F and Liu R S 2019 *Catal. Today* **335** 395–401
[50] Safardoust-Hojaghan H and Salavati-Niasari M 2017 *J. Clean. Prod.* **148** 31–6
[51] Moon J, An J, Sim U, Cho S P, Kang J H, Chung C, Seo J H, Lee J, Nam K T and Hong B H 2014 *Adv. Mater.* **26** 3501–5
[52] Briscoe J, Marinovic A, Sevilla M, Dunn S and Titirici M 2015 *Angew. Chem. Int. Ed.* **54** 4463–8
[53] Duan J, Zhao Y, He B and Tang Q 2018 *Electrochim. Acta* **278** 204–9
[54] Hu C, Li M, Qiu J and Sun Y P 2019 *Chem. Soc. Rev.* **48** 2315–37

[55] Bera D, Qian L, Tseng T K and Holloway P H 2010 *Materials* **3** 2260–345
[56] Vercelli B 2021 The role of carbon quantum dots in organic photovoltaics: a short overview *Coatings* **11** 232
[57] Li, L, Li M, Liang J, Yang X, Luo M, Ji L, Guo Y, Zhang H, Tang N and Wang X 2019 *ACS Appl. Mater. Interfaces* **11** 22621–7
[58] Wei G, Zhao X, Du K, Huang Y, An C, Qiu S, Liu M, Liu M, Yao S and Wu Y 2018 *Electrochim. Acta* **283** 248–59
[59] Syed Zainol Abidin S N J, Mamat S, Rasyid S A, Zainal Z and Sulaiman Y 2018 *J. Polym. Sci.* A **56** 50–8
[60] Moghimian, S and Sangpour P 2020 *J. Appl. Electrochem.* **50** 71–9
[61] Prasath A, Athika M, Duraisamy E, Sharma A S, Devi V S and Elumalai P 2019 *ACS Omega* **4** 4943–54
[62] Shaker M, Riahifar R and Li Y 2020 *FlatChem.* **22** 100171
[63] Wei G, Zhao X, Du K, Wang Z, Liu M, Zhang S, Wang S, Zhang J and An C 2017 *Chem. Eng. J.* **326** 58–67
[64] Sahoo S, Satpati A K, Sahoo P K and Naik P D 2018 *ACS Omega* **3** 17936–46
[65] Ji Z, Liu K, Li N, Zhang H, Dai W, Shen X, Zhu G, Kong L and Yuan A 2020 *J. Colloid Interface Sci.* **579** 282–9
[66] Ji Z, Dai W, Zhang S, Wang G, Shen X, Liu K, Zhu G, Kong L and Zhu J 2020 *Adv. Powder Technol.* **31** 632–8
[67] Wang J, Fang Z, Li T, Rehman S, Luo Q, Chen P, Hu L, Zhang Q and Wang H 2019 *Adv. Mater. Interfaces* **6** 1900049
[68] Ji Z, Li N, Xie M, Shen X, Dai W, Liu K, Xu K and Zhu G 2020 *Electrochim. Acta* **334** 135632
[69] Zhang X, Wang J, Liu J, Wu J and Chen H 2017 *Carbon* **115** 134–46
[70] Narayanan R 2017 *J. Solid State Chem.* **253** 103–12
[71] Li L, Li M, Liang J, Yang X, Luo M, Ji L, Guo Y, Zhang H, Tang N and Wang X 2019 *ACS Appl. Mater. Interfaces* **11** 22621–7
[72] Syed Zainol Abidin S N J, Mamat S, Rasyid S A, Zainal Z and Sulaiman Y 2018 *J. Polym. Sci., Part A: Polym. Chem.* **56** 50–8
[73] Moghimian S and Sangpour P 2020 *J. Appl. Electrochem.* **50** 71–9
[74] Zhang M, Liu W, Liang R, Tjandra R and Yu A 2019 *Sustain. Energy Fuels* **3** 2499–508
[75] Huang Y, Shi T, Zhong Y, Cheng S, Jiang S, Chen C, Liao G and Tang Z 2018 *Electrochim. Acta* **269** 45–54
[76] Wei G, Du K, Zhao X, Wang Z, Liu M, Li C, Wang H, An C and Xing W 2017 *Nano Res.* **10** 3005–17
[77] Moon J, An J, Sim U, Cho S P, Kang J H, Chung C, Seo J H, Lee J, Nam K T and Hong B H 2014 *Adv. Mater.* **26** 3501–5
[78] Liu W, Yan X, Chen J, Feng Y and Xue Q 2013 *Nanoscale* **5** 6053–62
[79] Lim S Y, Shen W and Gao Z 2015 *Chem. Soc. Rev.* **44** 362–81
[80] Islam M S, Deng Y, Tong L, Roy A K, Faisal S N, Hassan M, Minett A I and Gomes V G 2017 *Mater. Today Commun.* **10** 112–9
[81] Kakvand P, Rahmanifar M S, ElKady M F, Pendashteh A, Kiani M A, Hashami M, Najafi M, Abbasi A, Mousavi M F and Kaner. R B 2016 *Nanotechnology* **27** 315401
[82] Zheng, L, Guan L, Song J and Zheng H 2019 *Appl. Surf. Sci.* **480** 727–37

[83] Frueh S J, Coons T P, Reutenauer J W, Gottlieb R, Kmetz M A and Suib S L 2018 *Ceram. Int.* **44** 15310–6
[84] Ma Y, Yuan W, Bai Y, Wu H and Cheng. L 2019 *Carbon* **154** 292–300
[85] Zhang S, Sui L, Dong H, He W, Dong L and Yu. L 2018 *ACS Appl. Mater. Interfaces* **10** 12983–91
[86] Fei H, Ye R, Ye G, Gong Y, Peng Z, Fan X, Samuel E L G, Ajayan P M and Tour J M 2014 *ACS Nano* **8** 10837–43
[87] Kundu S, Yadav R M, Narayanan T N, Shelke M V, Vajtai R, Ajayan P M and Pillai V K 2015 *Nanoscale* **7** 11515–9
[88] Qiu H, Sun X, An S, Lan D, Cui J, Zhang Y and He W 2020 *J. Colloid Interface Sci.* **567** 264–73
[89] Shukla S K, Kushwaha C S, Shukla A and Dubey G C 2018 *Mater. Sci. Eng.* C **90** 325–32
[90] Tripathi B P, Dubey N C, Subair R, Choudhury S and Stamm M 2016 *RSC Adv.* **6** 4448–57
[91] Wang S, Shen J, Wang Q, Fan Y, Li L, Zhang K, Yang L, Zhang W and Wang X 2019 *ACS Appl. Energy Mater.* **2** 1077–85
[92] Syed Zainol Abidin S N J, Mamat M S, Rasyid S A, Zainal Z and Sulaiman. Y 2018 *Electrochim. Acta* **261** 548–56
[93] Jia H, Cai Y, Lin J, Liang H, Qi J, Cao J, Feng J and Fei W 2018 *Adv. Sci.* **5** 1700887
[94] Tang J, Ge Y, Shen J and Ye M 2016 *Chem. Commun.* **52** 1509–12
[95] Wang P, Zhang Y, Yin Y, Fan L, Zhang N and Sun K 2018 *ACS Appl. Mater. Interfaces* **10** 11708–14
[96] Fong K D, Wang T and Smoukov Sustain S K 2017 *Energy Fuels* **1** 1857–74
[97] Qu G, Cheng J, Li X, Yuan D, Chen P, Chen X, Wang B and Peng H 2016 *Adv. Mater.* **28** 3646–52
[98] Meng Q, Cai K and Chen L 2017 *Chen Nano Energy* **36** 268–85
[99] Gu, J, Hu M J, Guo Q Q, Ding Z F, Sun X L and Yang J 2014 *RSC Adv.* **4** 50141–4
[100] Li X, Lau S P, Tang R, Ji P and Yang 2014 *Nanoscale* **6** 5323–8
[101] Acerce M, Akdoğan E K and Chhowalla M 2017 *Nature* **549** 370–3
[102] Zhu Y, Ji X, Pan C, Sun Q, Song W, Fang L, Chen Q and Banks C E 2013 *Energy Environ. Sci.* **6** 3665–75
[103] Hu Y, Zhao Y, Lu G, Chen N, Zhang Z, Li H, Shao H and Qu L 2013 *Nanotechnology* **24** 195401
[104] Zhu, Y, Wu Z, Jing M, Hou H, Yang Y, Zhang Y, Yang X, Song W, Jia X and Ji X 2015 *J. Mater. Chem.* A **3** 866–77
[105] Chen Q, Hu Y, Hu C, Cheng H, Zhang Z, Shao H and Qu L 2014 *Phys. Chem. Chem. Phys.* **16** 19307–13
[106] Lv L, Fan Y, Chen Q, Zhao Y, Hu Y, Zhang Z, Chen N and Qu L 2014 *Nanotechnology* **25** 235401
[107] Li Q, Cheng H, Wu X, Wang c f, Wu G and Chen S 2018 *J. Mater. Chem.* A **6** 14112–9

IOP Publishing

Nanomaterials for Advanced Energy and Power Storage Devices

K R V Subramanian, N Sriraam, T Nageswara Rao and Aravinda C L Rao

Chapter 4

Importance of redox active electrolyte for the next generation energy storage system

Gaytri, Partha Khanra, Pankaj Kumar and Sudesh Mittal

With drastic depletion of fossil fuels, the development of an effective energy storage device has been an important frontier research field. Recently, energy storage devices have been paving the way towards potential application from portable to heavy machinery systems. Therefore, advanced energy storage research is subdivided into three parts, i.e. (i) electrode, (ii) electrolyte, and (iii) separator membrane. Within these the role of the electrode and the separator membrane are regularly investigated. Additionally, research into electrolytes coincides with energy storage developments to increase the energy and power density. In general, to improve the energy density and power density the chronological development is ongoing by controlling the electrodes properties, whereas the easy transportation of electrons from electrode to electrolyte or vice-versa and intercalation of electrolyte have been the focus. In addition to this, the redox active electrolyte has been established as another solution for improvement of energy density of energy storage systems. This chapter reviews the type of redox active electrolytes and focuses on the mechanism of operations for how to practically increase the energy density of the hybrid capacitor and redox flow battery based on a recently published paper.

4.1 Introduction

The development of an efficient electrochemical energy storage (EES) system for portable and heavy electronic machinery is the primary concern for the scientific research community, due to the rapid depletion of fossil fuel and the increase of global warming caused by carbon dioxide (CO_2) emissions from fossil fuel combustion [1, 2]. However, renewable energy sources, such as solar, wind, geothermal, biomass, and hydropower are available, whereas the discharge of pollution is very minimal compared to fossil fuel burning, but these resources are intermittent and depend on the natural calamity [3–5]. Therefore, to achieve continual energy

doi:10.1088/978-0-7503-4901-7ch4

production and its proper storage for one time applications, effective EES devices are being developed in various research centres by the modification of electrode and electrolyte materials. In general, the mechanism of EES is based on major two types, (i) batteries, where redox reaction is conducted at the interface of electrode and electrolyte, and (ii) electrochemical capacitors (EC) which are based on a double-layer electrochemical capacitor (EDLC) [6, 7]. The storage mechanism, working principle, cyclic stability, energy storage capability, and power deliverable capacity are dissimilar in these two types of devices [8–10]. Recently, advanced energy storage systems have been developed, whereas both batteries and EC working principles are assimilating each other. Among the EES system, batteries are the most widely used devices, powering a broad range of electronics from portable to heavy machinery. These include modern electronic heavy vehicle systems, where energy storage mechanisms are based on Faradic reactions or redox reactions, which have been conducted either at the negative terminal [11] or positive terminals [12] or in both electrodes [13] by intercalation/insertion or depositions of cations or anions, respectively. It is needed to know that anions are exchanging electrons in the negative electrode and cations are in the positive electrode. This mechanism provides a stable operating voltage window and high specific charge, but suffers from longer charging time and poor cyclic life [12–14]. This is because, during intercalation/de-intercalation of ions, the crystal structures of the electrode materials deteriorate, lowering the cyclic life, increasing the charging time, and lowering the charge storage capacity during continuous usage [14]. For these reasons, batteries are incapable of long time application. Furthermore, in lithium battery applications, the dendritic formation of lithium causes batteries to explode [15]. In contrast, in the primary electrochemical energy storage systems like EDLCs, the charge storage mechanism is based on electrochemical double-layer formations on the surface of the electrodes, whereas opposite polarized ions are adsorbed by electrostatic interactions, without any electron exchange between the electrode and electrolyte. This results in high-power delivery and ultralong cyclic life [16–20]. For example, EC mechanisms deploy a cyclic life that is higher than >10 000, has a power deliverable capacity of >10 kW kg^{-1} and interestingly rapid charging process are attractive for many devices, with precise start-up/stop times of the electronic devices [20]. It should be noted that the concentration of physical adsorption or electrostatic adsorption (no electron transition) and de-adsorption of ions during the charging and discharging, respectively, are highly dependent on the effective surface area of the electrode, electrolytes wettability (depends on the surface functionality), and porosity of the electrode materials [19, 21, 22]. However, ECs suffer from high self-discharge and lower energy density, compared to batteries, but it has better calendar life. Most electronics devices demand constant power, however, to compensate for the continuously dropping voltage during discharge, an EC must supply increasing current. Once an EC has discharged to below 50% of its maximum operating voltage, the residual capacity is essentially impractical, due to limitations of the power electronics [23–25]. Additionally, the increasing IR drop could be associated with high currents to maintain constant power supply, resulting in the fact that ~25% theoretical energy of the electrochemical double-layer (EDL) based EC are not practically useful. A specific power and energy density graph of EDLC, batteries and hybrid redox electrochemical energy storage system is depicted in figure 4.1 with comparative C-rates.

From figure 4.1, it can be concluded that the redox active electrolyte has higher energy density and power density performance compared to EDLC and batteries, respectively. Therefore, redox active EC systems or hybrid capacitors have become the frontier research field, which has added both redox reaction as the battery action and electrostatic double-layer charge storage mechanism as the EC system, to store the electrochemical energy. In general, electrochemical double-layer (EDL) based EC systems suffer from lower energy density and higher self-discharge rate compared to batteries [21]. On the other hand, batteries suffer from lower power deliverable ability, due to the slow kinetics of the ion's disadsorption from electrodes [24, 25]. Apart from these things, batteries cannot be used fully for several applications, such as electrical vehicle systems and other start-up devices where high instant power are required [26].

In figure 4.2 various EC mechanisms are shown: (1) indicating the electrochemical double-layer formations, which indicate the capacitive actions; (2) intercalation type Faradic redox reaction, i.e. battery action; (3) surface redox reactions, which is also a battery type action; and (4) redox electrolyte, which is randomly used in redox flow batteries, for grid scale energy storage applications, in addition to redox-non-flow energy storage applications (either supercapacitor or batteries), which is described here [27].

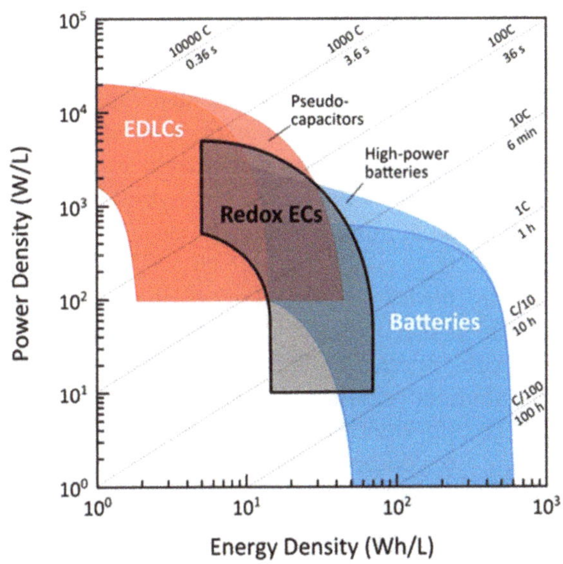

Figure 4.1. Ragone plot illustrating the range of device-level power and energy density for EDLCs, pseudocapacitors, secondary batteries, and redox-enhanced electrochemical capacitors (redox ECs). Diagonal lines correspond to different charge/discharge times and the corresponding C-rates. Reprinted with permission from [9]. Copyright (2017) American Chemical Society.

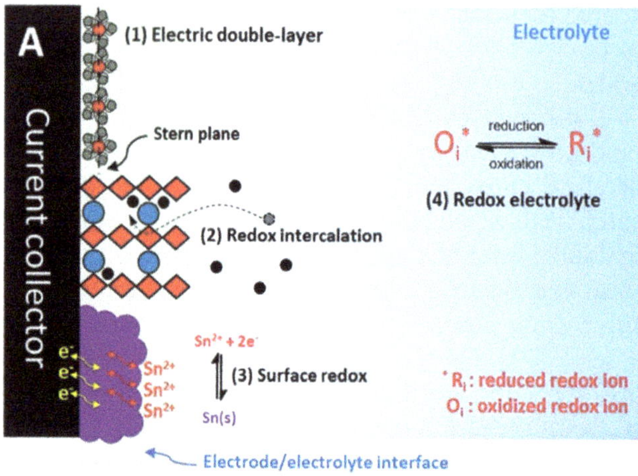

Figure 4.2. It is worth noting that, in the redox charge storage mechanism in the EC system both the electrode and electrolytes are taking part for faster electron transitions, whereas battery mechanism is incapable. Reproduced from [28]. CC BY 3.0.

To improve Faradic activity in the storage device, a great deal of research has been conducted for the improvement of redox activity of the electrode materials by changing the chemical compositions and its nano-structure. Therefore, different methodologies have been developed to synthesize the effective electrode materials. Alongside advanced electrode materials, electrolytes play a crucial role in developing efficient and hybrid electrochemical energy storage devices. They also help to mitigate the limitations of electrodes. For example, depending on the synthesis methods, the electrode materials suffer from lack of uniformity in pore-volume distributions [9, 24, 25, 29, 30], inefficient crystal structures [31] and several reacted residual molecules or impurities exist within the synthesized electrode materials, which shorten the electrode's life and working activities.

The advantage of redox active electrolytes use in hybrid ECs is the large amount of charge storage in small potential range or under the electrolyte's dissociation potential [9, 30, 32], redox reaction can be conducted with non-redox active electrode materials. Figure 4.3 indicates the measured capacitance value whereas EDL and redox contributions are shown in the coloured bars [28]. It demonstrates that increasing the concentration of redox species significantly enhances charge storage capacity, while the EDL contribution remains constant. In this study, an ion-exchange membrane was used; the importance of this component will be described in section 4.4. Notably, increasing the concentration of redox species dramatically improves capacitance performance within the same potential range 0.0–0.8 V. Obviously, the proper concentrations of redox active species with suitable solvents with respect to the electrode materials are highly desirable for the enhancement of cell performance. In general, appropriate redox-active electrolyte materials are (1) easily dissolvable into the electrolytic solvent and scalable, (2) easy to fabricate for commercial applications, (3) easy to avoid complex synthesis, and (4) high-power performances can be

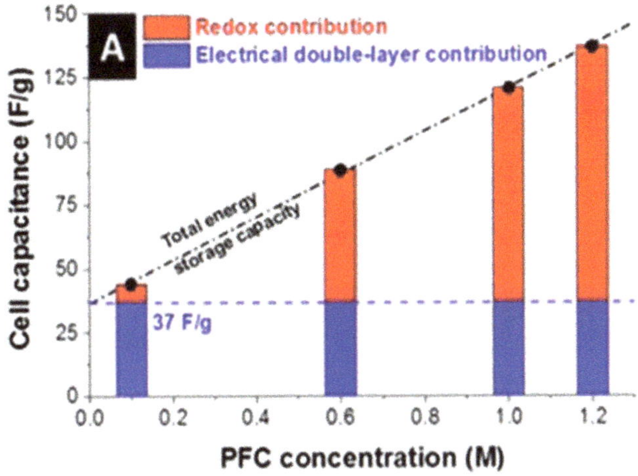

Figure 4.3. The contribution of redox species with increasing the concentration using ions exchange membrane. Reprinted with permission from [34]. Copyright (2016) American Chemical Society.

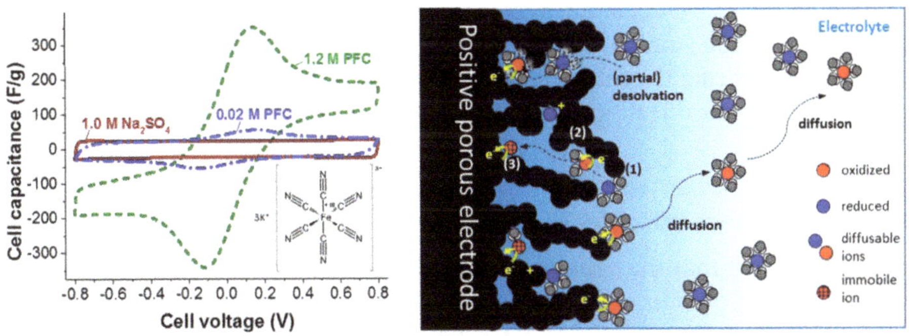

Figure 4.4. (A) Depiction of the CV graph of EDLC type in the presence of Na_2SO_4 (red) and the presence of different concentration (0.02 and 1.2 M) redox active electrolyte PEC (blue and green) reproduced from [30] CC BY 4.0, copyright The Author(s) 2015, published by ECS; (B) indicated the charge storage mechanism by redox active species.

maintained. The redox reactions execute over a large surface of electrode and equivalent charge can be store in positive and negative electrodes by the ion absorption and/or electron transition by redox species, which will be reversed during discharge from both electrodes. To get better performance chemical reactions or solid-state diffusion should not occur [24], and (5) by controlling redox-active organic molecular structures, one can tune the solubility and redox potential voltage [33].

Moreover, in figure 4.4, a schematic illustration of a redox-electrolytes based hybrid system is presented with effective cyclic-voltammetry (CV) performance in the absence and presence of redox species, in fact, the Na_2SO_4 was used as aqueous electrolyte and potassium ferricyanide (PFC) was used as redox electrolyte. In this study, Na_2SO_4 helped to store the charge by electrochemical double-layer formation and simultaneously the PFC took part for the redox reactions to improve the

Faradic charge storage. Therefore, the resulting charge storage was improved, which was indicated by the coverage area of the CV graph. The greater area covered by the CV graphs indicated more charge storage by the modified redox electrolyte arrangement [28].

In this context, the basic formula for the charge storage amount can be defined as follows $C = \frac{\Delta Q}{\Delta V} = \frac{I\Delta t}{\Delta U}$, whereas '$\Delta Q$' indicate the net store charge or capable of releasing the charge amount, 'ΔV' is the range of operating voltage and 'Δt' indicated the time difference to reduce the voltage from V_{initial} to V_{final}. But for the non-linear curve as shown in figure 4.4, the claimed capacitance is as follows [30]:

$$C_{\text{claimed}} = \frac{2i_m \int \Delta V.\, dt}{V^2 \mid \frac{V_i}{V_f}} \tag{4.1}$$

whereas, i_m indicates the mass normalizing current, and V_i and V_f are the initial and final voltage respectively. The determination of actual capacitance value by the EDLC and redox active process were described by Akinwolemiwa *et al* [30], which is out of the scope of this chapter. In this perspective, this chapter has discussed the advanced EES system, where the electrolytes play the dynamic role in improving the performances of the storage device. Additionally, pore-volume distributions play a vital role in enhancing the charge storage capacity and energy deliverable properties by reducing the ion diffusion paths [24, 25, 30], in a later section we will briefly described it.

4.1.1 Charge storage principles of redox electrolytes containing ECs

It was established that the capacitance of ECs was increased by implementing the redox activity or pseudocapacitive contribution during charging and discharging. Redox active electrodes or redox active species can contribute to the redox activity, which we have indicated in the above section, whereas PFC was used as a redox mediator. Similarly, Lota *et al* established the iodide/iodine redox pair with carbon electrode and highest specific capacitive performance achieved 1840 F g^{-1} [28]. Therefore, in redox electrolyte system, the soluble redox couples in electrolyte are taking part for the redox reactions at the electrode/electrolyte interface, which was described by Akinwolemiwa *et al* that has illustrated in figure 4.5 [30]. In figure 4.5, 'O' and 'R' simultaneously indicate the oxidized and reduced states of redox electrolytes. During charging, redox species (either O_{bulk} or R_{bulk}) primarily enter inside the pores by electrostatic force between the electrode and electrolytes. Processes 1 and 1′, represent the equilibria of the de-solvation and solvation states of redox species. Then the redox species achieved the transition state O* or R* by processes (2) and (2′), respectively, which are the transition states before the electron-transfer reaction with electrode materials. These transition states are progressed further and converted to adsorption R_{ads} and O_{ads} by the process (3) and (3′), i.e. not the standard electrode reaction, but effectively to increase the charge storage capacity for ECs system. Then electron-transfer is conducted by the adsorbed species via process (4) on the internal and external surface of the electrode.

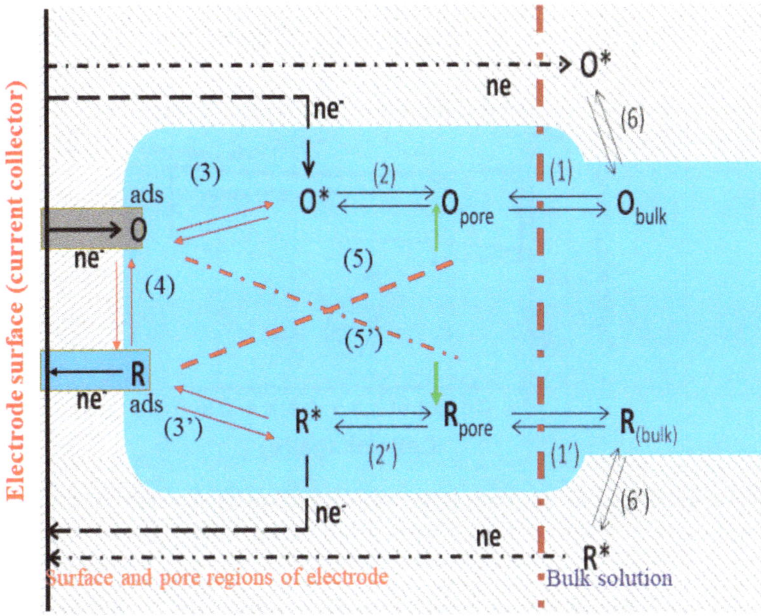

Figure 4.5. Redox mechanism of electrode.

Process (4) is primarily responsible for improving the charge storage capacity of the electrode. In process (4), electrons are transferred between the adsorbed redox species and the internal and external surfaces of the electrode, i.e. directly responsible for the enhancement of charge storage capacity of the device. The point to be noted is that, the transition state achieved by the redox species (O* and R*) may also undergo electron-transfer reaction by processes (5) and (5′), without participating in any adsorption of the transition state. At the same time as electron-transfer, the redox species R_{ads} and O_{ads} is converted to a soluble product R_{pore} and O_{pore}, which can diffuse through the pores and return again in the bulk electrolyte via processes (5) and (5′). Additionally, pores on the surface are also take part for the electron-transfer reaction via (6) or (6′) and similarly redox ions diffuse to the bulk electrode, but pores can restrict or shorten the ion diffusion from pores to bulk electrolyte [24, 30, 34, 35]. It is worth noting that the rate of redox reactions depends on the thermodynamic stability of the redox ions with the pH of the supporting electrolyte and redox potential, which can be depicted by Pourbaix diagram in figure 4.6 [24, 36].

In general, a redox-active electrolyte at the positive electrode (catholyte) is oxidized, on the other hand, redox-active electrolyte at the negative electrode (anolyte) is reduced [9]. Similarly, during discharging, the reverse processes occur. The resulting current moves in and out (during charging and discharging, respectively) of the storage system via current collectors. In the above discussion, the anolyte and catholyte are fully miscible throughout the cell volume and make a homogeneous state during the discharging of the cell. Interestingly, with binary pH

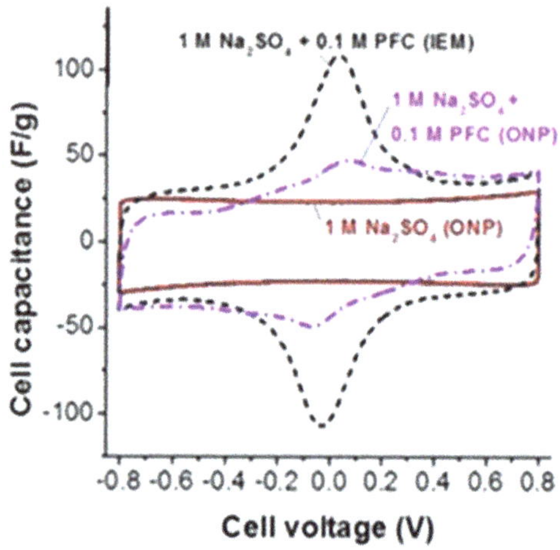

Figure 4.6. A comparative cyclic-voltammetry curve with ion-exchange membrane and porous membrane in redox electrolyte system. Reprinted with permission from [28]. Copyright (2016) American Chemical Society.

catholyte and anolyte-based redox electrolytes are reported for the battery type energy storage system, which are effective in improving the charge storage capacity and suppressing the self-discharge of the cell [37]. The point to be noted is that redox electrolyte strategy is commonly used in redox flow batteries, whereas ion-exchange membranes (IEM) are used to protect the migration of the 'liquid-redox-active' electrolyte. In the sense of the electric charge storage process, the redox active EC system is similar to the redox flow battery mechanism. However, the non-flow redox-EC approach does not currently apply to manufacture, because of manufacturing difficulties and cost issues. In spite of these difficulties, several research outcomes are reported to the betterment of energy storage strategies, which is shown in figure 4.5(c) [28]. Moreover, in redox active systems, the ion-exchange membranes insulate the miscibility of catholyte and anolyte, but maintain the ionic conductivity [9]. The greater importance will be described in section 4.4. But, from the above discussion it can be concluded that the porosity and kinetics of redox ions diffusions play the major role for efficient cell operations, which will be described briefly in section 4.1.2.

4.1.2 Mechanism of diffusion, adsorption, and kinetics of redox electrolytes

It is well known that the high surface-to-volume ratio electrode materials are highly demanding for energy storage applications, which facilitate the increase in the contact area of electrode and electrolytes [29, 38, 39]. Additionally with high surface area, the porosity has unique behaviour to enhance the cell performance, such as faster electron transportation, high storage capacity, easy charging and discharging, etc. It is worth mentioning that, electrolytes form a thin film on the surface of the electrode, which indicates the double-layer formation and from this thin layer electrons are exchanging during redox reactions [21, 24, 25, 30]. Therefore, fast charge–discharge processes and power deliverable properties can be controlled by the electrolyte's kinetics which are highly controlled by the surface structure and porosity of the electrodes. Additionally, the surface functional groups also control the electrolyte kinetics [22], which is not described in this chapter. This thin layer electron exchange or thin layer electrokinetics (TLE) theory was first described in the 1960s by Hubbard and Anson [39]. They showed that electrochemical efficiency is highly controlled by the diffusion length of electrolytes or molecular dynamics. Later, Narayanan, Bandaru, and coworkers clarified the high-power deliverable capacities by molecular dynamics that depend on porous structure of the electrode [25]. The molecular dynamics in porous structures can be defined by surface diffusion, Brownian motions, and Knudsen diffusion [40]. Therefore, by optimizing pore diameter or the spacing of electrode material layers, the diffusional limitations can be mitigated, resulting in high yield charge–discharge rates that can be achieved with attractive cycling stability. For example, in the presence of activated carbon (AC) as an electrode material, dual redox active electrolyte materials such as KI and $VOSO_4$ were used as anolyte and catholyte, respectively. From this study, it was observed that the charge storage capacity was dramatically decreasing with increasing the discharge rates or high scan rate [41, 42]. By computational modelling study of Faradaic process in uniform porous electrodes was ascribed to the reduced capacity, perhaps due to planar diffusion from the exterior surface but not within the pores of the electrode. For better performance, the pore structure or nano-spacing should be smaller than the equivalent diffusion layer thickness (δ) [25]:

$$\delta \sim \sqrt{\pi D t} \sim \sqrt{\frac{\pi D \Delta V}{s}} \tag{4.2}$$

where 'D' indicates the ion diffusion coefficient, ΔV is the corresponding voltage range and 's' is the scan rate during operation. From equation (4.2), it can be predicted that at low scan rate 'δ' should be large and overlap with planar diffusion limited current (i_{diff}). Moreover, the diffusion layer thickness is also proportionally controlled by the projected area of the electrode (A_{porj}), bulk concentration of redox active species (n_0), and the $s^{1/2}$. Moreover, due to the thin layer formation and entrapment of the ions within the pores of the micro-electrode a thin layer electrokinetics (TLE)-based current would be generated (i_{TLE}) that should be proportional to the (V_{pore}) scan rate (s), which is shown in equation (4.4). At the medium scan rate, the i_{diff} will occur from the top of the surface; therefore, the concentration gradient will be generated between the surface (which is in direct contact with the bulk electrolyte) and near pore-volume (whereas the redox species are entrapped), resulting additional i_{diff} will be governed. Simultaneously, the effects of i_{TLE} will be reduced, resulting in that diffusional current being dominant. At the higher scan rate the diffusional current will dominate which is proportional to the effective area (A_{eff}) of the electrode, due to a complete decoupling of the diffusion layer. Finally, the amount of charge storage provided by the solvated redox-species is given by

$$Q = xFV_{\mathrm{pore}}n_0 \tag{4.3}$$

i.e. entrapped in the V_{pore} in the TLE regime. Where 'x' denotes the number of electrons transferred to electrode per redox reaction. Accordingly, high electrochemical energy storage capability and rate capability may be controlled in the TLE regime via designing an electrode with high porosity and small 'δ'. Due to the Faradaic reactions, the redox species are homogeneously depleted/accreted, therefore, all the charges are controlling. Furthermore, greater kinetics can be achieved by the absence of diffusion, then the charge storage capacity would be directly proportional to the confined volume (V_{pore}) of electrolyte in all of the pores and its bulk concentration (n_0). The corresponding current can be derived by accompanying the TLE based peak currents (i_{TLE}) [25]

$$i_{\text{TLE}} = \frac{F^2}{4RT}(sn_0 V_{\text{pore}}) \tag{4.4}$$

Equation (4.4) indicates that the thin layer electrochemical current depends on the scan rate of the device. The peak current (i_p) of the cell can be defined as the sum of electrochemical double-layer current i_{EDL}, diffusional current i_{diff}, and thin layer electrochemical current i_{TLE} [24, 25]

$$i_p = i_{\text{EDL}} + i_{\text{diff}} + i_{\text{TLE}} \tag{4.5}$$

At the higher redox-activity, the i_{diff} and i_{TLE} is higher than i_{EDL}. Finally, it can be concluded that the uniform high porous structure is highly desirable for the improvement of energy storage, which is also reported in many published papers [43].

4.2 Categories of redox electrolyte

From the above discussion, it can be concluded that electrochemical energy storage could be improved by redox active species. This chapter has focused on various electrolytes, precisely redox based electrolytes, which have been reporting in various published papers for hybrid capacitor (HC) applications. Moreover, the modified redox electrolytes are used in grid level battery type energy storage systems. It should be noted that the redox reactions can be classified into two basic mechanisms, i.e. (i) Faradic adsorption of solute ions near the surface of the materials, or (ii) underpotential deposition of redox elements. The Faradic adsorptions are basically surface controlled rather than diffusion controlled, which facilitates faster redox reactions, for example, for hydrous RuO_2 in aqueous acidic solution, a proton (H^+) can be adsorbed Faradaically near the surface of the electrode [24, 30]. In contrast, in underpotential deposition mechanisms, two kinds of metal salts are used, whereas one type of metal is deposited on a second metal. In this process less energy (underpotential) is required than the redox potential of the first metal for deposition. Additionally, this redox mechanism is helpful to increase the cyclic life of the device, such as deposition of lead on a gold surface.

In summary, the total charge storage mechanism of hybrid electrochemical capacitor (HEC) can be categorized in four different mechanisms in between electrode and electrolyte, i.e. (i) electrostatic double-layer (EDL) formation, (ii) bulk redox reaction of the electrode (electrodes are active participants for the redox reactions), (iii) redox reaction near the electrode surface, (both electrode and electrolyte can take part), and (iv) redox activity of the bulk electrolyte (the electrolytes are active for charge storage, by changing the oxidation state) [24].

In general, the charge storage capacity of a porous electrode material in a redox electrolyte can be described by the simple expression in equation (4.6) [24, 25],

$$Q_T = q_{dl} + q_{\text{Electrolyte}} + q_{\text{Electrode}} \tag{4.6}$$

where Q_T is the total charge stored in the electrode, q_{dl} represents the storage charge due to EDL formation, and $q_{\text{Electrolyte}}$ and $q_{\text{Electrode}}$ correspond to the charge stored due to redox reactions between electrode and electrolyte, respectively. Remarkably, $q_{\text{Electrode}}$ and $q_{\text{Electrolyte}}$ completely depend on the applied potential, which works in a limited region of potential, i.e. oxidation and reduction potential of redox elements [24, 25, 34]. Furthermore, the redox potential also depends on the experimental condition, pH of the medium composition, including the temperature of the electrolyte and the composition of the electrode materials. The point to be noted is that the total charge formations highly depends on the ionic mobility and conductivity redox active molecules. For the ionic conductivity three types of electrolytes are reported, i.e. aqueous electrolyte, non-aqueous electrolyte, and polymer electrolyte, which help the ion's conduction during charging and discharging. The point to be noted is that the ionic mobility and conductivity of redox active molecule's liquid electrolyte are higher than polymer electrolyte. Moreover, the conductivity of the particular species (*m*) can be defined as [24, 25, 34]

$$\sigma = \sum_m n_m.\ \mu_m.e.\ z_m \tag{4.7}$$

where, 'μ_m' is the ionic mobility, 'n_m' is the concentration of charge carriers, 'e 'is the elementary charge, and 'z_m' is the magnitude of valence of the mobile ion charges. The variables of equation (4.7), are lean on solvation effect (i.e. depend on pH), migration of the solvated ion (i.e. temperature and viscosity), and lattice energy of the solute molecules or salt. Thus, all the parameters including solvents, additives, and salts can affect the conductivity of the electrolytes, obviously concentration of additive and salt should be adjusted to better kinetics.

On the basis of the pH, the electrolyte can be classified as acidic, basic, and neutral electrolyte. However, the nature of the solvent can be classified as (i) aqueous based redox electrolytes, (ii) organic based redox electrolytes, and (iii) IL-based redox electrolytes, which is described below. In earlier reports, Senthilkumar *et al* differentiated the redox electrolyte systems by way of (i) *redox additive liquid electrolytes*, (ii) *redox active liquid electrolytes*, or (iii) *redox additive-polymer electrolyte* [13]. In a redox additive electrolyte, the redox compounds are directly added with electrolyte. For example, hydroquinone can be added with H_2SO_4 electrolyte, whereas hydroquinone is redox additive. Similarly, potassium ferricyanide aqueous 1 M KOH solution has sufficient ionic (anion and cation/$[Fe\ (CN)_6]^{3-}/3K^+$) conductivity and performed as reversible redox reactions, which has been reported by Su *et al* [44]. In this study, the potassium ferricyanide ($K_4Fe\ (CN)_6$) acted as a *redox-additive* compound.

In contrast, redox active electrolytes are directly involved in charge-transfer reactions, no additional redox species are involved or added in this system. This type of electrolyte can perform both as a solvent and a redox mediator also, no supporting electrolyte such as Na_2SO_4, KOH, H_2SO_4, etc are involved. The operation pathways are very simple and are economically safer, firstly the ions are migrated from bulk electrolyte and adsorbed on the electrode surface. Then the redox reaction or the electron transition is involved between the electrode and electrolyte interface, simultaneously the redox mediators form the electrochemical double layer on the surface of the electrode. $K_4Fe(CN)_6$ can act as a redox active electrolyte in absence of KOH, because $Fe(CN)_6^{4-}$ ions can be converted into Fe $(CN)_6^{3-}$ during charging and reverse it during discharging [34]. These redox reactions are conducted on the positive electrode. Therefore, in this redox mediator system the positive electrode is contributing more to storing electrochemical energy. Similarly, KI can be used as a redox active electrolyte, which was reported by Lota *et al* whereas polyiodide formed during redox reactions on the positive electrode. The specific capacitance value is shown to be higher than H_2SO_4 in the presence of an active carbon electrode [33]. In this type of arrangement, electrochemical energy storage density is higher than the negative electrode. Later, Frackowiak *et al* introduced a novel method, which could help to remove the imbalance of charge storage density in both the positive and negative electrodes. In that study, KI and $VOSO_4$ were used as redox active mediators, whereas the reversible oxidation–reduction redox peaks were observed. In the positive electrode iodine ions were oxidized in the positive electrode. On the other hand, VO^{2+}-ions were reduced by the oxygenated compounds on the surface of the electrode by exchanging protons from oxygenated compounds on the surface of the carbon electrode [33]. Similar studies were reported by many researchers in different electrochemical environments, such as gel-polymer electrolyte medium. The gel-polymer electrolytes are attractive for commercial applications, because they are less corrosive and more flexible than aqueous based electrolytes. However, they suffer from poor electrolyte wettability, which hinders the ionic accessibility, resulting in lower capacitive performance [13]. Therefore, Faradic reaction or pseudocapacitive performance is required to improve the capacitive performance [45]. The redox additive gel-polymer electrolyte was reported by Yu *et al*, where PVA-KOH gel-polymer electrolyte was developed and KI was added as redox active mediator. It was observed that with increasing KI concentration, the capacitance value was improved and simultaneously reduced the equivalent series resistance value. In summary, during the charging in redox electrolytes based EES systems catholyte ions are oxidized at positive electrodes and anolytes ions are reduced at negative electrodes. Moreover, during discharge the reverse phenomena are conducted. But it was observed that the oxidation and reduction potential are different, which leads to higher self-discharging. Therefore, the catholyte and anolyte-based EES systems are deployed in recent reports. Evamko *et al* showed various examples of catholyte and anolyte-based energy storage systems with various redox species [9]. Additionally redox based hybrid capacitor, redox flow batteries have been developed for grid level energy storage systems, whereas, the redox species diffuse to bulk electrolytes after exchanging the electrons during charging. As a result, the oxidation rate of bulk electrolytes changes and it then flows to the separate reservoir by a pumping system. In this system, large scale energy can be stored, but it suffers from lower energy density and power density as compared to conventional batteries. Later, many reports have been published with different chemical environments with improved capacitive performance, which have been discussed and summarized below in the light of charge properties.

With reference to redox-electrolytes, another satisfactory classification was reported by Lee *et al* in a review paper, whereas redox species were classified into the following three categories [24]:

(i) anionic redox electrolytes (e.g., ferrocyanide/ferricyanide or negatively charged ions)
(ii) cationic redox electrolytes (e.g., Cu^+/Cu^{2+} or positively charged ions)
(iii) neutral redox electrolytes or (Non-ionic redox electrolytes) (e.g., quinone/hydroquinone or not ionized in any charged before applying voltage).

In the following sections classified redox species with disperse mediums (solvent) are discussed, redox species are dissolved in various solvent, i.e. aqueous, gel-polymer or solid-polymer, and ionic liquid-based solvents. The total performance of EES systems depends on the selection of proper electrolytes in accordance with electrode materials. Initially, various aqueous based redox–electrolyte systems are described with supporting electrolyte in section 4.2.1. The pH of the supporting electrolyte could be neutral or basic or acidic.

4.2.1 Aqueous systems

4.2.1.1 Anion redox electrolytes

Precisely, anionic redox electrolytes are based on negatively charged molecules or groups of molecules, such as halides (Br^-, I^- etc) [46–49], pseudo-halides such as inorganic thiocyanate [50], organometallic complexes such as ferricyanide and ferrocyanide [51–54], and organic compounds [55] such as indigo carmine have been successfully used in various published reports. Halide, bromine, and iodine-based salts have been investigated more, and particularly bromide (Br^-) and iodide (I^-) have a high reversible redox-potential and small ionic-size facilitating fast diffusion kinetics in aqueous electrolytes. Moreover, bromide salts are inexpensive, abundant, and have an excellent theoretical capacitance value of 334.97 mAh g^{-1} [56–58]. Noticeably, during the electrochemical charging process, the Br^--ions loses electrons on the anode and is oxidized to bromine (Br^0). But in solution bromine usually combines with free bromide ions and forms Br_3^-. Similarly, during the discharge reverse reactions occur in negative electrodes. The charging–discharging reaction in the presence of bromide-based redox couples are as follows:

during the charging process:

$$2Br^- - 2e^- \rightarrow Br_2 \tag{4.8}$$

$$Br_2 + Br^- \rightarrow Br_3^- \tag{4.9}$$

during the discharging process:

$$Br_2 + 2e^- \rightarrow 2Br \quad (4.10)$$

and standard electrode potential $E_0 = 1.08$ V versus SHE [59]. In general, alkaline bromide salts are dissolved in aqueous solvent and carbon-based materials are used as electrode materials. In the presence of a bromide-based redox couple, both EDLC and pseudocapacitive (battery type) behaviour combine to increase the electrochemical energy storage. Li *et al* have shown that at lower (>1.6 V) operating potential, the charge storage mechanism depends on the EDLCs, but above 1.6 V operating potential the bromide ions participate in redox reactions [60]. Particularly, at higher potential the bromide (Br^-) ions will diffuse from the bulk electrolyte to the pores of carbon and pass through an intermediate state (Br*) which oxidize and transfer the electron to the electrode and generate Br_3^-. The redox process has been described in section 4.1.1. Moreover, it has been observed that with increasing operating voltage the specific capacitance value and energy density of the system have improved due to the redox reactions at higher operating potential. Interestingly, Tang *et al* have shown that, when Na_2SO_4(1 M) was added with KBr (0.5 M), the capacitance value improved dramatically [61]. As compared with Na_2SO_4-KBr redox couple, $ZnSO_4$-KBr redox-couple has also been observed by Yu *et al* [62] and they observed that the power density was dramatically improved over a conventional capacitor. In this study, the Br^--anions and Zn^+-cations both work at the same time on cathode and anode, respectively. Therefore, both electrodes store charge by redox reactions which is higher than EDLC mechanism. Addition of alkaline bromide, transition metal bromide ($FeBr_3$) has also been studied in the presence of H_2SO_4 and Na_2SO_4 electrolytes [63]. The foremost shortcoming of bromide-based redox-electrolyte is the cross-diffusion of bromide ions that reduce the capacitive performance of the device. To avoid the cross-diffusion, several additional salts have been added, such as NH_4Cl, KCl, K_2SO_4, bromine complex etc have been used in various reports [64–66]. The organic salt, 1-ethyl-3-methylimidazolium bromide was dissolved in ionic liquid 1-ethyl-3-methylimidazolium tetrafluoroborate and provided a stable charging–discharging performance. By addition of bromide ions, the capacitive performance was doubled and the rate capability was also reducing [65, 66].

The mechanism of electrochemical charge storage of an iodine-based redox-couple is analogous to the bromide ions. But iodine can provide a higher capacitance value (310 mAh g^{-1}) than bromide (211 mAh g^{-1}) [67]. It is important to note that, the iodine-based redox system highly depends on the pH of the solution, basically the iodine redox pair formed in neutral and acidic solution is I^-/I_3^-, and in a basic solution is I_2/IO_3^- [68, 69]. The standard redox potential of I^-/I_3^- and I_2/IO_3^- are 0.54 and 0.697 V, respectively, with respect to normal hydrogen electrode (NHE) [33, 67]. The ionic size of iodine is higher than bromide ions but less than 2 nm, that facilitates shuttle from bulk electrolyte to the micropores (<2 nm) and mesopores (2–50 nm) of the electrode materials. Interestingly, the capacitance performance of iodide-based redox species also depends on the ionic radii alkaline metal halide such as LiI, NaI, KI, RbI, CsI, etc [33, 70]. The alkaline iodide salt was dissolved and a study of the electrochemical performance in the presence of an activated carbon electrode was done by Lota *et al.* From this study, it was observed that the energy storage performance increased with increasing ionic radius of the alkaline metal cation. The observed storage performance was 300, 492, 1078, 2272, and 373 F g^{-1} in LiI, NaI, KI, RbI, and CsI, respectively [70]. However, the ionic radius of Cs^+ is higher than other alkaline metal cations, but showed lower electrochemical energy storage capacity. Svensson *et al* reported that the iodide's structure particularly depends on the nature and dimension of the counter ion, herein a alkali metal cation [71]. In the experiments of Lota *et al*, during Faradic reactions polyiodides, such as I^{3-}, I^{5-} or even I^{9-} were formed in iodide solution on the positive electrode. Probably, polyiodides are more thermodynamically favourable than the simple I^-. Moreover, in the presence of Li^+, Na^+, K^+, and Rb^+ polyiodides are linearly formed and fit for the micro/macro pore of the carbon structure [70], which is easy for diffusion, mobility, etc. The rubidium iodide-based electrolyte showed the highest capacitive performance and better reversibility of ions than other alkaline halides, because smaller polarizing power and conductivity of ions are inversely proportional to ionic radii. Therefore, I^- ions are more attracted to positive electrodes and metallic ions are defuse to the electrolytes. In contrast, the caesium iodide showed lower capacitance, because polyiodides are formed in a non-symmetrical, non-linear and curved manner, that hinders the easy migration of ions. Jiang *et al* showed that a KI based redox-electrolyte is also useful for the metal based redox electrode. In this study, mesoporous MnO_2 was used as the electrode material and Na_2SO_4 and KI was used the electrolyte separately. It was observed that, in the presence of Na_2SO_4, a regular CV curve appeared. In contrast, in the presence of a KI electrolyte with different concentration, the resulting CV curves were not symmetrical and the calculated energy storage capacity was higher than Na_2SO_4. Additionally, with increasing concentration of KI, the energy storage capacity was improved. In a similar manner, Senthilkumar *et al* reported an asymmetric supercapacitor, whereas flower-like α-Bi_2O_3 and bio-derived activated carbon were used as the negative and positive electrodes, respectively [73]. A comparative electrochemical performance was observed in the presence of KI and/or Li_2SO_4 as the electrolyte. In this study, it was observed that, in the presence of a Li_2SO_4 electrolyte the asymmetric supercapacitor arrangement showed the redox peaks in the CV curve, but with the addition of KI with this arrangement, the CV curves showed a better capacitive area than the Li_2SO_4 electrolyte. In comparison with KBr, KI shows better performance for energy storage. However, iodide and bromide-based electrolytes are also going through the same process, i.e. polyiodides (I^{3-}, I^{5-} or even I^{n-}) and polybromide (Br^{3-}, Br^{5-} or even Br^{n-}) in positive electrode [33, 56, 67–73]. The addition of halide redox electrolyte, pseudo-halide (group of atoms which consist with two or more electronegative atoms and at least one of them is nitrogen) based compounds showed admirable performances, such as aqueous solution of metallic and ammonium based thiocyanates (KSCN, NaSCN, LiSCN, and NH_4SCN) provided better capacitive properties, because of adsorption of SCN^- ions by positive electrode during charging, then oxidized by transferring electron to electrode ($SCN^- \rightarrow SCN + e$) and finally polythiocyanogen formation on the surface of the electrode. These steps are common in anion redox electrolytes [50]. It is worth noting that, halide or pseudo-halide-based redox species offer single electrons to the electrode during charging. But in Faradic-based energy storage systems, charge storage density increases with increasing numbers of electron transferred from redox species to electrodes. On that account, sulfur based anion redox electrolyte has attracted great attention in the energy storage industry, because sulfur ions can transfer two electrons during Faradic reactions (S^{2-} of $S^0 + 2e^-$) [74–77]. Moreover, the theoretical capacitance value of sulfur ions is higher than that of halide-based compounds, i.e. 1674.87 mAh g^{-1}, however, the standard redox potential of sulfide based redox species is −0.45 V versus SHE, which is lower than bromide-based species. In this context, Qian *et al* for the first time showed sulfide based redox electrolyte for hybrid energy storage applications. The CuS nanotube was used as the electrode and a hybrid electrolyte was prepared using 0.5 M NaOH, 0.5 M Na_2S, and 0.5 M sulfur powder. In this study, the CV curves appeared with several redox peaks and offered higher capacitance values as compare to pure NaOH electrolyte [74]. From figure 4.6(b), it can be concluded that the peak position of O3/R3 (O—oxidation/R—reduction) appeared due to Faradic reaction between CuS and OH^- molecules in following way:

$$CuS + OH^- \leftrightharpoons CuSOH + e^- \tag{4.11}$$

In contrast, the presence of polysulfide electrolytes, an additional two redox peaks appeared (O1/R1) and (O2/R2) in figure 4.6(b), which indicated redox reactions by polysulfide molecules in following ways:

$$S^{2-} \leftrightharpoons S^0 + 2e^- \tag{4.12}$$

$$S^{2-} \leftrightharpoons (x/8) \quad S_8 + 2e^- \tag{4.13}$$

$$S_x^{2-} + 2e^- \leftrightharpoons S_{x-1}^{2-} + S^{2-} \tag{4.14}$$

$$S_8 + 2e^- \leftrightharpoons S^{2-} \tag{4.15}$$

The O1 peak appeared due to oxidation of S^{2-} and produced S^0 (S atom) (equation (4.12)), which was easily dissolved in concentrated Na_2S and formed polysulfide anions (S_x^{2-}) by the following equation:

$$(x - 1)\, S^0 + S^{2-} \rightarrow S_x^{2-} \tag{4.16}$$

At higher potential the polysulfide anions are converted to elemental sulfur by oxidation (equation (4.13)) and appeared as oxidation peak O_2. The elemental sulfur deposited on the surface of the CuS electrode. During discharge, two reduction peaks appeared, which indicated R2 and R1. The R2 peak appeared due to reduction as shown in equation (4.15). Another peak R1 appeared due to the reduction from polysulfide anions S_x^{2-} to S^{2-} as equation (4.14). However, the complete redox reaction of sulfur based electrolytes is incompletely explored [74]. The addition of sulfide anions, the persulfate anions based electrolyte has also been explored by Han *et al* [78]. In this study, KOH (2 M) and KOH (2 M) + $Na_2S_2O_8$(0.04 M) was used as the redox electrolyte. Noticeably, no significant change was observed in CV curves, but the galvanometric charge–discharge graph was slightly increased in the presence of $Na_2S_2O_8$. Table 4.1 lists many anion-based redox electrolytes and their performance.

Analogous to pseudo-halides, organometallic based redox electrolytes have also reported as anion based redox electrolytes, which are efficient for reversible redox reaction, during charging and discharging. The organometallic based redox couple is hexacyanoferrate ($K_4Fe(CN)_6/K_3Fe(CN)_6$) [28, 44, 79]. Su *et al* showed that when potassium ferri- and ferrocyanide were dissolved in basic (4 M KOH) solution, the electrochemical performance enhanced two times that of pure KOH electrolyte in the presence of a NiO electrode [44]. Later, Lee *et al* showed the excellent electrochemical performance of PFC in the presence of a Na_2SO_4 electrolyte, whereas active carbon was used as the electrode materials and a porous membrane was used to control the ion's diffusion. In this study, it was observed that with increasing the concentration of PFC, the electrochemical performance was improved and increased by applying an ion-exchange membrane, the capacitive performance dramatically increased and the self-discharge rate reduced [28]. From the above discussion, it can be concluded that anion based redox species have paramount value.

4.2.1.2 Cationic redox electrolytes

In the previous sections, various anion based redox electrolytes or negatively charged electrolytes and their working principles have been discussed. In this section cation-based redox-couples and their working principles will be described. Basically, the aqueous cationic redox electrolytes are based on multivalent transition metal (e.g., Fe, Cu, V, etc) and also lanthanides (e.g., Ce) have to be found as cation-based redox electrolyte [24, 30, 42, 80]. $Ce_2(SO_4)_3$ salt was used as cation redox species in H_2SO_4 electrolyte medium and electrochemical performance was moderately increased. The major challenge of lanthanide based redox system is its redox potential range, which is higher than HER and OER potential [80]. Apart from these metal cations, the organic molecules like methylene blue and viologen are also effective for the cation-based redox reactions [81, 82]. In general, cations do not easily take part in electro-sorption on a positive electrode and similarly anions are not attracted to the negative electrode. Therefore, cation affinitive groups are attached to the electrode surface which facilitate chemisorption of cations. In general, the carbon atoms are functionalized by oxygenated compounds (–OH, –COOH, etc), due more to the high edge site reactivity than the inside of the solid carbon [42]. These groups are fully or partially deprotonated during charging, resulting in the surface being negatively charged. The addition of electronegative, oxygen, nitrogen sulfur and other halogen groups also have the tendency to attract metal cations. Therefore, high surface area carbon surfaces are effective for cation mediated redox reactions in hybrid capacitors. To develop cation-based redox electrolytes, metal salts are dissolved in neutral or pH based aqueous electrolyte, for example, $CuSO_4$-$FeSO_4$-H_2SO_4 [83], $VOSO_4$-H_2SO_4, and $CuCl_2$-HNO_3 [84] are tested with functionalized carbon surfaces. Practically, it was observed that charge storage is greatly dependent on the porosity of electrode materials and also

Table 4.1. Redox mediated anionic aqueous electrolyte.

Electrolyte	Electrode configuration	EES type	Capacity	Potential window (V)	Capacity stability (%)
Organic anionic					
Indigo carmine, H_2SO_4	MWCNT|MWCNT	Battery-like	50 F g^{-1}	0/1	−30% (10 000 cycles, 360 mA g^{-1})
Organometallic complex					
$K_3Fe(CN)_6$, KOH	Co-Al-LDH, half-cell	Battery-like	712 F g^{-1}	−0.1/0.5 (SCE)	−33% (200 cycles, 2 A g^{-1}) [7]
$K_3Fe(CN)_6$, KOH	M_xCl_y-CB25), half-.cell	Battery-like	12.7 F cm^{-1} [2, 14]	−0.1/0.45 (SCE)	—
$K_3Fe(CN)_6$, KOH	MC-CF|CuO-CWT	Capacitor-like	269 mF cm^{-1} [2, 5]	0/1.5	+48% (2000 cycles,4 mA cm^{-1} [2, 7]
$K_3Fe(CN)_6$, KOH	MC|Ni-Co-O-RGO	Capacitor-like	126 F g^{-1} [5]	0/1.5	−40.2 (3000 cycles, 2.5 A g^{-1})
$K_3Fe(CN)_6$, KCl	CNT, half-cell	Battery-like	28 mAh cm^{-2}	−0.6/0.6 (SCE)	−10% (5000 cycles, 0.2 A cm^{-2}) [7]
$K_3Fe(CN)_6$, Na_2SO_4	Fe_2O_3|MWCNTs-MnO_2	Capacitor-like	226 F g^{-1} [9]	0/2	−17% (500 cycles, 200 mV s^{-1})
$K_3Fe(CN)_6$	MC|MC	Capacitor-like	304 mAh g^{-1}	0/1.8	−20% (9000 cycles, 1 A g^{-1})
$K_3Fe(CN)_6$, $K_4Fe(CN)_6$, KOH	NiO, half-cell	Battery-like	156 F g^{-1}	−0.1/0.55 (SCE)	−3.6% (1000 cycles, 2 A g^{-1}, 20 °C) [7]
$K_3Fe(CN)_6$, $K_4Fe(CN)_6$, Na_2SO_4	DN|DN	Battery-like	73 mF cm^{-1} [2, 7, 14]	0/2.4	0% (5 mA cm^{-2})
PPD [9], KOH|$K_3Fe(CN)_6$, KOH	AC-CFP|$Co(OH)_2$-GNS [17]	Battery-like	205 F $g^{-5)}$	0/2.0	0% (20 000 cycles, 10 A g^{-1})
$K_4Fe(CN)_6$, KOH	Co-Al-LDH, half-cell	Battery-like	304 F g^{-1}	−0.1/0.5 (SCE)	−6% (200 cycles, 2 A g^{-1}) [7]

$K_4Fe(CN)_6$	MC∣MC	Capacitor-like	272 F g^{-1}	0/1.2	—
$K_4Fe(CN)_6$, H_2SO_4	PANICB	Capacitor-like	912 F g^{-1}	0/1	0% (100 cycles, 5 mA cm^{-2})
Halides					
MgI_2	Zn∣poly(2,5-thienylene)	Battery-like	—	—	—
ZnI_2	Iodine-6-nylon, half-cell	Battery-like	—	−0.2/0.5 (Ag/AgCl)	—
ZnI_2	Zn∣iodine-6-nylon	Battery-like	—	—	—
ZnI_2	Zn∣iodine-6-nylon	Battery-like	—	—	−23% (300 cycles, 2 mA cm^{-2})
ZnI_2	Zn∣polypyrrole-I_2	Battery-like	—	—	—
ZnI_2	Zn∣ferrocene-DCA compound[23]	Battery-like	800 mAh g^{-1} [24]	—	—
ZnI_2	Zn∣MC	Battery-like	64 mAh ml^{-1} [12]	—	14% (500 cycles, 2.26 mA cm^{-2})
ZnI_2, NH_4I	Zn∣polyaniline	Battery-like	143 mAh g^{-1}	—	—
ZnI_2, NH_4I	Zn∣iodine-6-nylon-CB-CF	Battery-like	60 mAh ml^{-1} [12]	—	—
KI	MC∣MC	Capacitor-like	300 F g^{-1}	0/0.8	+28% (10 000 cycles, 1 A g^{-1})
KI	MC∣MC	Capacitor-like	234 F g^{-1}	0/0.8	—
KI	MC∣MC	Capacitor-like	126 F g^{-1} [5]	0/1.6	—

(Continued)

Table 4.1. (*Continued*)

Electrolyte	Electrode configuration	EES type	Capacity	Potential window (V)	Capacity stability (%)
KI	MC|MC	Capacitor-like	493 F g^{-1}	0/1.4	+4% (10 000 cycles, 1 A g^{-1})
KI, H_2SO_4	PANI-MWCNTs|MWCNTs	Battery-like	726 F g^{-1} [5]	0/1	−3% (500 cycles, 1 A g^{-1})
KI, H_2SO_4	MC|MC	Battery-like	912 F g^{-1}	−0.2/0.8	+26% (4000 cycles, 5 mA cm^{-2})
KI, Na_2SO_4	MC|MC	Battery-like	604 F g^{-1}	−0.2/0.8	—
Ki, Li_2SO_4	A-Bi2O3|MC	Battery-like	99.5 F g^{-1} [5]	0/1.6	−28% (1000 cycles,4 mA cm^{-2})
KI, KOH	RGO-CB|RGO-CB	Capacitor-like	500 F g^{-1}	0/0.8	−14% (5000 cycles, 8.3 A g^{-1})
KI, Na_2MoO_4, H_2SO_4	MC|MC	Battery-like	470 F g^{-1} [5]	0/1	−35% (5000 cycles, 8 A g^{-1})
VOSO4|KI	MC|MC	Capacitor-like	500 F g^{-5})	0/0.8	—
LiI	MC|MC	Capacitor-like	178 F g^{-1}	0/0.8	—
NaI	MC|MC	Capacitor-like	203 F g^{-1}	0/0.8	—
RbI	MC|MC	Capacitor-like	220 F g^{-1}	0/0.8	—
CsI	MC|MC	Capacitor-like	234 F g^{-1}	0/0.8	—
$ZnBr_2$	Zn|polypyrrole-Br_2	Battery-like	—	—	72% (capacity,16 h)

$ZnBr_2$, NH_4Br	Zn|polyaniline	Battery-like	100 mAh g^{-1}	—	—
$ZnBr_2$, NH_4Cl	CC|Polyacrylamide-MC	Battery-like	—	—	—
KBr, H_2SO_4	MC|MC	Battery-like	572 F g^{-1}	−0.2/0.8	—
KBr, methyl viologen dichloride [18]	MC|MC	Battery-like	49.8, 13.3 mAh g^{-1} [10]	0/1.4	—
KBr, heptyl viologen dichloride [19]	MC|MC	Battery-like	44.0, 12.1 mAh g^{-1} [10]	0/1.4	—
NaBr, pentyl viologen dibromide [20]	MC|MC	Battery-like	—	0/1.2	−3% (10 000 cycles, 2.5 A g^{-1})
Ethyl viologen dibromide21), H_2SO_4	MC|MC	Capacitor-like	408 F g^{-1} [9]	0/0.9	+30% (1000 cycles, 2.5 A g^{-1})
EVBr, Tetrabutylammonium bromide, NaBr	MC|MC	Battery-like	—	0/1.35	−10% (5000 cycles, 2 A g^{-1})
Pseudohallide					
KSCN	CB|CB	Capacitor-like	—	0/1.6	−31% (10 000 cycles,-)
NaSCN	MC|MC	Capacitor-like	64 F g^{-1}	0/1.4	—
LiSCN	MC|MC	Capacitor-like	97 F g^{-1}	0/1.6	—
NH4SCN	MC|MC	Capacitor-like	43 F g^{-1}	0/1.2	−13% (10 000 cycles,-)

(Continued)

Table 4.1. (*Continued*)

Electrolyte	Electrode configuration	EES type	Capacity	Potential window (V)	Capacity stability (%)
Redox surfactants					
TPAB, KOH	MC-CB|MC-CB	Capacitor-like	161 F g^{-1}	—	−33% (5000 cycles, 1 A g^{-1})
TPAI, KOH	MC-CB|MC-CB	Capacitor-like	165 F g^{-1}	—	−5% (5000 cycles, 1 A g^{-1})
TPAI, Na_2SO_4	MC|MC	Capacitor-like	81 F g^{-1}	0/0.6	—
Sodium lignosulfonates, $H_2SO_4^-$	MC-CB|MC-CB	Capacitor-like	181 F g^{-1} [13]	0/0.8	—

concentration of cations (Fe+, Cu+, VO^{2+}, etc). Many published reports are shown in table 4.2. Interestingly, in the presence of single cations such as Cu^{2+} the reversible reaction was lowered, due to irreversible depositions of Cu-cations on the surface of the electrode. However, after the addition of Fe-cations, the re-oxidation and desorption happened in suitable potential ranges. The complete reaction mechanisms are shown equations (4.17) and (4.18) [30].

During charging in positive electrode:

$$Fe^{2+} + Cu = Fe^{3+} + Cu^{2+} + 3e^{-} \tag{4.17}$$

During discharging in negative electrode:

$$xFe^{3+} + yCu^{2+} + zC + (x + 2y)e^{-} = (Fe_xCu_y)ads^{2x+}C_z \tag{4.18}$$

Precisely, metal salts are dissolved in acidic solutions and act as redox additive electrolytes. Remember that this mechanism is based on electrodeposition techniques, whereas, the metal cations are converted to metal on the surface of positive electrodes. Therefore, the single cation-based electrolytes are not suitable for reverse reactions and it was observed that double metal cation-based electrolytes are rendering reversible reactions. Similarly, it was observed that, when $CuCl_2$ was used in the presence of HNO_3, the charge storage amount was increased dramatically even though the concentration of $CuCl_2$ was very low. In this study, good reversible charging–discharging at high current loading was indicated. The phenomenon can be explained by electrosorption and chemisorption of the Cu-cations in the presence of the surface carbonyl groups [26]. Furthermore, in the presence of $VOSO_4$ in H_2SO_4 solution produced vanadyl cation $[VO(H_2O)_5]^{2+}$, commonly abbreviated as VO^{2+}. In this study, from the bulk electrolyte the VO^{2+} cations were deported to proton-exchange carbonyl groups or oxy-functional groups on the surface of negative carbon electrodes. Finally, the –COOVO + bond was formed between the oxy-functional group of carbon electrode and the VO ion [42]. The capacitance value in this system showed 670 F g^{-1}. The foremost shortcoming in this system was back diffusion of VO_2-cations to the bulk electrolyte, resulting in high self-discharge. Therefore, to prevent this mitigation of VO_2-cations, the redox conjugated KI was added [42]. By these methods, the self-discharge of the redox capacitor can be suppressed. Many reports were published with different transition metal cations added in acidic mediums. Therefore, many metallic salts are studied as redox additives in electrolytes such as Zn^{2+}, Sn^{2+}, Mn^{2+}, Fe^{2+}, Ni^{2+}, etc, which operates in low electrode potential. However, these electrodeposition type redox reactions are unable to change the oxidation state of the electrolyte effectively. Interestingly, it was observed that if metal halide is used as a redox electrolyte then effective results are observed. Lee *et al* showed that zinc iodide is used as a redox electrolyte whereas activated nanoporous carbon was used as an electrode material, resulting the system exhibiting low self-discharge rates without applying any ion-exchange membrane and being capable of high-power deliverable capability [85]. The major reason for higher diffusion kinetics was confinement of the redox ions in nanopores, which led to thin layer diffusion rather than planner diffusion. Similar phenomena are also observed in SnF_2 coupled with VOSO4 [27, 86].

Table 4.2. Redox mediated cationic aqueous electrolyte.

Electrolyte	Electrode configuration	EES type	Capacity	Potential window (V)	Capacity stability (%)
Transition metals					
Fe^{2+}, Cu^{2+}, H_2SO_4	MC-GP∣MC-GP	Battery-like	223 mAh g^{-1}	—	−26% (400 cycles, 0.5 A g^{-1})
$VOSO^4$∣KI	MC∣MC	Capacitor-like	500 F g^{-1} [5]	0/0.8	—
$VOSO_4$, H_2SO_4	MC, half-cell	Battery-like	2.1 F cm^{-2}	0.2/1.3 (Ag/AgCl)	−8% (5000 cycles, 0.1 A g^{-1})
$VOSO_4$, $SnSO_4$, H_2SO_4	MC∣MC	Battery-like	—	0/1.4	−15% (4500 cycles, 1 A g^{-1})
SnF_2, H_2SO_4∣$VOSO_4$, H_2SO_4	MC∣MC	Battery-like	—	0/1.4	−20% (6500 cycles, 1 A g^{-1})
$CuCl_2$, HNO_3	PCMS∣PCMS	Capacitor-like	294 F g^{-5}	0/1.35	−0.9% (1000 cycles)
$(NH_4)_2Fe(SO_4)_2$, H_2SO_4	MC-GP	Battery-like	1499 F g^{-1}	−0.5/0.5	−6.2% (10 000 cycles, 0.1 A g^{-1})
Lanthanide					
$Ce_2(SO_4)_3$, H_2SO_4	MGP∣MWCNT	Battery-like	64 F g^{-1}	0/1.5	−6% (3000 cycles, 44 mA cm^{-2})
Organic cations					
Methylene blue [17], H_2SO_4	MWCNT∣MWCNT	Battery-like	23 F g^{-1} [5]	0/1	—
Methylene blue [17], H_2SO_4	MWCNT∣MWCNT	Capacitor-like	14 F g^{-1} [5]	0/1	—
Methyl viologen dichloride [18], KBr	MC∣MC	Battery-like	49.8, 13.3 [10]	0/1.4	—
Heptyl viologen dichloride [19], KBr	MC∣MC	Battery-like	44.0, 12.1 [10]	0/1.4	—
Pentyl viologen dibromide [20], NaBr	MC∣MC	Battery-like	—	0/1.2	−3% (10 000 cycles, 2.5 A g^{-1})
Ethyl viologen dibromide [21], H_2SO_4	MC∣MC	Capacitor-like	408 [9]	0/0.9	+30% (1000 cycles, 2.5 A g^{-1})
EVBr, tetrabutylammonium bromide, NaBr	MC∣MC	Battery-like	—	0/1.35	−10% (5000 cycles, 2 A g^{-1})

It is necessary to mentioned that in aqueous electrolytes hydrogen and oxygen evolution reaction was observed, because of higher reduction potential of transition metal cations. It was observed that when ion selective membranes for transition metal redox electrolytes were used, the leakage current was suppressed, resulting in slow self-discharge rate. In this context Chun *et al*, Evanko *et al* and later, many redox combinations were reviewed, whereas redox additives with different pH are separated by selective ion-exchange membranes only to suppress the self-discharge via redox shuttling [9].

With the addition of metal cations, the organic compound-based redox species were also revealed. Among them, methylene blue (MB) as a redox active electrolyte, has shown excellent reversible redox reaction and enhanced the capacitance performance 4.5 times that of pure H_2SO_4 electrolyte. From the galvanostatic graph, it was observed that MB redox reaction was developed in both the multi-wall carbon nanotube (MWCNT) electrode and voltage range is 0–0.104 V. Therefore, the resultant energy storage performance was increased dramatically [81]. Similarly, 1,1-diethyl-4,4-bipyridinium bromide (viologen) redox active electrolyte was used in the presence of H_2SO_4 (1 M) to improve the electrochemical energy density. Interestingly, viologen ionized in bromine-ion and 1,1-diethyl-4,4-bipyridinium ions. During the electrochemical performance, the bromine was oxidized on positive electrodes and 1,1-diethyl-4,4-bipyridinium ions reduced on negative electrodes. Therefore, a synergic redox behaviour was observed, which helped to enhance the electrochemical performance of activated charcoal [82]. Moreover, viologen generates di-cations, i.e. 1,10-dimethyl-4,40-bipyridinium cation (MV^{2+}), which could be strongly attracted by negative electrodes. But after reduction on the negative electrode MV^{2+} transformed to MV^{+}, which is also attracted by the negative electrodes, resulting in reduction of free diffusion and self-discharge rate. Therefore, in future, more study is required to improve the energy storage performance and reduction of self-discharge.

4.2.1.3 Neutral redox electrolytes or non-ionic redox electrolytes

In addition, with cation and anion based redox electrolytes, non-ionic redox electrolytes have come to the forefront of research, because they are non-flammable and non-corrosive compared to other electrolytes. Additionally, hydrogen or oxygen evolution rates are more competitive and depend on the degree of pH of the electrolytes [24]. In contrast, neutral electrolytes were developed by inexpensive inorganic salts as supporting electrolytes and provided higher safety features with environmental benevolence. It was observed that very few neutral redox species were reported previously, such as hydroquinone (HQ) [87, 88–91], catechol [88], rutin [92], water soluble conducting polymer p-nitroaniline (PNA) [93], sulfonated polyaniline (SPANI) [94], etc. The performance of neutral redox electrolytes is summarized in table 4.3. To alleviate the corrosion of electrode materials and other supportive parts, and simultaneously to extrude other undesired additional reactions, KCl, NaCl, KNO_3, KOH, H_2SO_4, etc were used as the supporting salts, which facilitated improving the ionic conductivity of the electrolytes. The major shortcoming of the neutral redox electrolyte was found to be redox shuttling, resulting in lower

Table 4.3. Redox mediated neutral aqueous electrolyte.

Electrolyte	Electrode configuration	EES type	Capacity	Potential window (V)	Capacity stability (%)
HQ, H_2SO_4	MC|MC	—	901 F g^{-1}	0/1	−65% (4000 cycles,4.42 mA cm^{-2})
HQ, H_2SO_4	MC|MC	Capacitor-like	220 F g^{-1} [5]	0/1	—
HQ, H_2SO_4	RGO-PANI| RGO-PANI	Battery-like	553 F g^{-1}	0/0.7	−36% (50 000 cycles, 10 A g^{-1})
HQ, H_2SO_4	PANI-CB|PANICB	Battery-like	773 F g^{-1}	0/0.7	−48% (10 000 cycles, 10 A g^{-1})
HQ, H_2SO_4	CNT, half-cell	Battery-like	2250 F g^{-1} [13]	0/1 (SCE)	−50% (500 cycles, 5 mV s^{-1})
HQ, H_2SO_4	PANI-SnO_2 [8]	Capacitor-like	857 F g^{-1}	−0.5/0.5	−17.8% (2000 cycles, 0.5 A g^{-1})
HQ, H_2SO_4	GHG|GHG	Capacitor-like	88.5 F g^{-1} [5]	0/0.8	
HQ, BQ, H_2SO_4	PEB|PEB	Battery-like	2646 F g^{-1}	0/0.65	+15% (50 000 cycles,12.5 mA cm^{-2})
AQDS, KNO_3	MC|MC	Capacitor-like	225 F g^{-1}	0/1.8	
Brominated hydroquinones, KOH	MC-CB|MC-CB	Capacitor-like	314 F g^{-1}	0/1.1	−25% (5000 cycles, 2 A g^{-1})
Rutin, H_2SO_4	MC|MC	Capacitor-like	86 F g^{-1} [5]	0/1	−4.4% (5000 cycles, 10 A g^{-1})
Pyrocatechol, H_2SO_4	MC|MC	Capacitor-like	136 F g^{-1} [5]	0/1	−3.6% (5000 cycles,10 A g^{-1})

Pyrocatechol, H_2SO_4	NMCCB|NMCCB	Battery-like	628 F g^{-1}	0/1	+13.5% (6000 cycles, 5 A g^{-1})
Catechol, H_2SO_4	PANIRGO| PANIRGO	Battery-like	409 F g^{-1} [5]	0/1.2	−10.3% (5000 cycles, 10 A g^{-1}) [7]
PPD, KOH	MnO_2|MnO_2	Battery-like	325 F g^{-1} [13]	−0.5/0.5	−25 (5000 cycles, 1 A g^{-1})
PPD, KOH	MnO_2-CB, half-cell	Capacitor-like	156 F g^{-1}, −20 °C	−0.4/0.55 (SCE)	−16% (600 cycles, 2 A g^{-1}, −20 °C)
PPD, KOH	MC-GP|MC-GP	Capacitor-like	605 F g^{-1} [13]	0/1	−5.5% (4000 cycles,-)
PPD, KOH	MC-GP, half-cell	Battery-like	852.3 F g^{-1}	−1/0 (SCE)	−29% (5000 cycles, 40 A g^{-1})
PPD, KOH	NC-TEG [17], half-cell	Battery-like	636 F g^{-1}	−1/0 (Ag/AgCl)	−12.6% (10 000 cycles, –)
PPD, KOH|$K_3Fe(CN)_6$, KOH	AC-CFP|Co $(OH)_2$-GNS	Battery-like	205 F g^{-1} [5]	0/2.0	0% (20 000 cycles, 10 A g^{-1})
MPD, KOH	MC-GP|MC-GP	Battery-like	78 F g^{-1} [5, 13]	−0.5/0.5	−9.3% (10 000 cycles, 1 A g^{-1})
PPD, H_2SO_4	MC-CB|MC-CB	Battery-like	121 F g^{-1} [5], −18 °C	0/1.5	—
Soluble conducting polymer					
SPANI, H2SO4	GHG|GHG	Capacitor-like	64 F g^{-5}),13)	0/0.8	−1% (1000 cycles, 1 A g^{-1})
p-Nitroaniline, KOH	Carbon-DMG, half-cell	Battery-like	386 F g^{-1}	−1/0 (SCE)	+13% (5000 cycles, 10 A g^{-1})

Coulombic efficiency (<90%), and high self-discharge rate, that was prevented by applying ion-exchange membrane. Therefore, in comparison with other hybrid energy storage systems, neutral electrolyte-based energy storage consisted of an ion-exchange membrane, which may be anion exchange membrane (AEM), proton-exchange membrane (PEM), cation exchange membrane (CEM), and semipermeable membrane (SPM).

Chen *et al* demonstrated the HQ-based hybrid energy storage system, where H_2SO_4 was used as ion conducting electrolyte and PEM was used as separator [95]. From this study, it was observed that redox shuttling was greatly suppressed after applying the Nafion 117 PEM membrane. Similarly, water soluble sulfonated polyaniline (SPANI) redox electrolyte was also studied using semipermeable membrane, which selectively allowed migration of SO_4^{2-} and H^+ like ions. The major shortcomings of soluble redox active polymer electrolyte were lower solubility. It was observed that charge storage capacity was controlled by concentration of redox active species in solvent. Therefore, more research is required in future to developed highly soluble polymers for efficient energy storage system.

4.3 Non-aqueous redox electrolyte system

4.3.1 Ionic liquid-based redox electrolytes

In general, the electrolytes are based on solvents and salt. The solvents are inert to store the electrochemical charge, and only take part to transport the ions from bulk electrolyte to electrode, i.e. ion conducting medium. In the cell, solvents are acquired in most of the pores of the electrode volume and solutes or salt concentrations are adjusted to achieve the maximum output. However, the solvent molecules are not taking part in electrochemical energy storage, whereas salt molecules are absent. Therefore, the performances of the cells are reduced. In the previous section we have mostly described the aqueous based electrolytes, which operate at low potential windows. In contrast, ionic liquid-(ILs) based electrolytes can take part in energy storage systems directly with above 3.5 V operating voltage window, no other additional salts are required [21, 96–98]. Additionally, The ILs based electrolytes are useful in overcoming the energy density lagging, because it is composed of organic cations and a weakly coordinating inorganic/organic anion [21, 24, 98–102]. Therefore, ILs are taking part in energy storage directly without any external salts or ion generating compounds. Moreover, since, in recent decades, the ILs have been acknowledged as excellent electrolytes because of their various outstanding properties, which benefited in formulating an effective ES-system, for example, high thermal stability, good chemical stability, and low corrosivity, tuneable structures for a large range of operating temperatures, high electrochemical operating voltage window (4–6 V) with excellent ionic conductivity at room temperature, i.e. in the range 10^{-3}–10^{-2} S cm^{-1}, and non-flammability with negligible vapour pressure at room temperature [96, 103–108]. The importance of redox active IL-based electrolytes is summarized below [109]:

(i) The high density of IL-ions: the electrochemical performance of ES-device could be increased by the increase of ions in the electrolyte medium. In the presence of redox active molecules with IL medium, the total ion density would be increased, which facilitate improvement of the storage density of the device. Furthermore, no other solvents are required to dissolve the redox active materials. Both IL and redox active components will take part for transferring charge to electrode that improve the performance of the device.

(ii) The wide stable electrochemical potential window: the operating electrochemical potential window is higher than the aqueous system. Therefore, higher redox potential redox-active components can be easily added with it, and facilitated to wider operating potential. In contrast, aqueous electrolyte is easily decomposed at higher potential and produced hydrogen and oxygen.

(iii) The excellent adaptability with redox active components: due to the unique structure and designability, the ILs can be tuned with specific redox active species to reduce the free diffusion and self-discharge

As per chemical composition, ILs can be differentiate in three type (1) protic, (2) aprotic,(3) zwitterionic, and according to the structure it can separated into three types (1) pyridinium, (2) imidazolium, and (3) quaternary ammonium [109]. No doubt, the redox based ILs could play the pioneering role to enhance the performance of energy storage devices, but the interaction with electrode needs to be improved more in future. However, anion or cation modifications of ILs facilitate to improve the interaction with the electrode, but sometimes it suffers from high viscosity. In general, the redox-active ILs are developing in following three ways:

(1) The additional redox active components are added with ILs, such as $CuCl_2$ added with 1-ethyl-3-methylimidazolium tetrafluoroborate ([Emim][BF_4]). In this methods sometimes viscosity can increased.

(2) The modifications of ILs by other redox active species, the anion or cation parts of the ILs can be modified by other redox active species, such as bis (trifluoromethanesulfonyl)imide anions with ferrocene. These methods are effective for removing the insoluble problems of some redox-active components, but ion size will be increased dramatically, which is not suitable for nanoporous or sometimes microporous samples.

(3) ILs, directly added with gel-polymer and redox active salts, such as PVA-Li_2SO_4-[Bmim]I, have been reported to generate flexible energy storage devices. They have strong elongation properties and are suitable as flexible energy storage devices.

Similar to aqueous systems, cation and anion based redox species can be developed, which are described below. The organic cations are imidazolium (Im), pyridinium (PY), pyrrolidinium (PYR), ammonium, and sulfonium and inorganic or organic anions are BF_4^-, PF_6^-, triflate ($CF_3SO_3^-$), and bis (trifluoromethanesulfonyl imide) (TFSI) (($(CF_3SO_2)_2N-$) etc. With the different combinations of cations and anions multiple possibilities of IL combinations can architectured for the desired applications. Imidazolium containing IL cations are broadly studied due to their high ionic conductivities and relatively low viscosities, which is effective for the fabrication of energy storage devices [24]. Many published reports have indicated that the ILs can be used directly as electrolytes and can also be used with organic solvents. Recently, for further improvement of energy density, redox additives are applied with ILs. Whereas, anion, cation and neutral based redox active materials have been added. Among the ILs, the 1-ethyl-3-methylimidazolium tetrafluoroborate ([Emim][BF4]) as an aprotic IL (AIL) has received special attention because of its wide operating potential window, low viscosity, good chemical stability, and excellent ion conductivity. In the absence of any other redox active salts it can help to store charge by developing double-layer capacitance. To enhance the energy storage performance, Sun *et al* added copper chloride (0.36 M) cations with the ILs, that increased the capacitance performance by changing the oxidation state (Cu(II)/Cu(I)/Cu(s)) [97]. Noticeably, from scanning electron microscopic (SEM) analysis after electrochemical performance the dendrite structure of copper has been observed on the surface of carbon electrode, which is a serious issue for an energy storage device.

Yamazaki *et al* showed the strong anion based redox additive 1-ethyl-3-methylimidazolium bromide ([EMIm]Br) with ionic liquid ([EMIm]BF4), resulting in a significant improvement of capacity from 18 to 33 mAh g^{-1}. In addition, several beneficial actions were observed, such as, low leakage current, low level redox shuttling and high Coulombic efficiency. Similarly metal cations, such as copper chloride (0.36 M) can be dissolved in 1-ethyl-3-methylimidazolium tetrafluoroborate ([EMIm]BF4), that increase the capacitance performance by changing redox oxidation state (Cu(II)/Cu(I)/Cu(s)) by dissolution of copper chloride (0.36 M). Similarly, neutral quinone based redox additives are effective for the enhancement of capacitance value. For the state-of-art protic ILs, triethylammonium bis(trifluoromethane)sulfonimide (TEATFSI) is employed as an electrolyte for activated charcoal (AC) and 0.3 M HQ has been dissolved for redox shuttling resulting in the enhancement of capacitance value (72 F g^{-1}) and an operating voltage window of nearly 2.5 V. It has been observed that the Coulombic efficiency in halide-based redox electrolytes is higher than cation-based and quinone-based compounds, due to strong adsorption of bromine atoms with the carbon electrode, which prevents diffusion of oxidized bromide to negative electrode. A similar phenomenon is also observed in boron-based redox moieties in aqueous based electrolytes.

In addition to these, another approach has been demonstrated by Xie *et al* with two different ILs, which were developed by modification of imidazolium cation and the bis(trifluoromethanesulfonyl)imide anion with ferrocene redox active moieties. These fuctionalized ILs showed large maximum operating voltage and suppress self-discharge with reversible oxidation of $[FcNTf]^-$ anions to $[Fc + NTf^-]^0$ at 1.7 V with respect to Pt electrodes. In general, single redox electrolyte redox reactions were conducted in either positive electrodes or negative electrodes, therefore, a potential imbalance developed, which triggered rapid potential flow on the opposite electrode. The potential imbalance could be overcome by adding binary (anions and cations) redox moieties. Mourad *et al* introduced a bi-redox IL for effective charge balance, whereas IL perfluorosulfonate (PFS^-) anion and imidazolium cation (MIM^+) were functionalized with anthraquinone (AQ) and 2,2,6,6-tetramethylpiperidinyl-1-oxyl (TEMPO) moieties, respectively. During charging, AQ-PFS^- adsorbed at the negative electrode and reduced to AQ^{-}-PFS^{-} and later on AQ^{2-}-PFS^{-}. On the other hand, MIM^{+}-TEMPO adsorbed at the positive electrode, then oxidized as MIM^{+}-$TEMPO^{+}$. Similar individual work had was conducted using pure bi-redox IL as a redox active electrolyte above room temperature (at 60 °C) without supporting 1-Butyl-3-methylimidazolium bis(trifluoromethylsulfonyl)imide [BMIm]TFSI like ions, which had exhibited capacitance value 370 F g^{-1}. In contrast, by using the [BMIm]TFSI with 0.5 M bi-redox IL showed lower capacitance value.

Until now, far less study has been conducted on ILs based system as compared to aqueous system.

4.4 Redox mediated gel-polymer electrolytes

In order to address the leakage concerns associated with liquid-based electrolytes and corrosion electrodes, polymer-based GEL electrolytes were developed in addition to aqueous and non-aqueous solvents. Good mechanical and chemical stability, good flexibility, easy redox moiety diffusion, great ionic conductivity liquids—higher than solid polymer electrolytes—and good thermodynamic stability might all be obtained by adding a gel-polymer-based electrolyte. Gel-polymer electrolytes, on the other hand, have reduced ionic fluidity and ionic dynamics in contrast to liquid-based electrolytes, making it impossible to carry out the quick adsorption/desorption of ions on the interface. Consequently, research has been done on the addition of redox species to the electrolyte in order to further increase capacitance performance. Simultaneously, the redox active species in the gel-polymer electrolytes restrict leak current and self-discharge by not hopping between the two electrodes. Furthermore, because GPEs have a low electrical conductivity on their own, it is critical to increase the gel electrolyte's conductivity in order to promote electron transport.

Gel electrolytes demonstrate the durability and flexibility of solids as well as the ease of liquid diffusion. They also allow for the flexible manufacturing of electronic devices and the easy movement of ions and specific materials between liquid and solid states [110, 111]. It is a potential electrolyte because of its many benefits, which include increased ionic conductivity compared to solid electrolytes, strong mechanical and chemical stability, etc [111–114].

A unique redox-mediated approach for SCs was recently disclosed, and it effectively increases the ionic conductivity and generates extra capacitance through the redox mediator's rapid, reversible redox reaction [115].

Indigo carmine (IC) [116], 2-mercaptopyridine (PySH) [117], 1-butyl-3-methylimidazolium iodide (BMIMI) [118], alizarin red S (ARS) [119], FeBr3 [120], 1,4 naphthoquinone, 1-anthraquinone, sulfonic acid sodium (AQQS) [121], and 1-ethyl-3-methylimidazolium tetrafluoroborate ([EMIM]BF_4) [122] are among the redox additives commonly found in gel-polymer electrolytes. Polyvinyl alcohol (PVA) and sulfuric acid (H_2SO_4) were combined with indigo carmine (IC) to create the redox-mediated gel-polymer electrolyte (PVA-H_2SO_4-IC). It gains 188% more ionic conductivity, or 20.27 mS cm^{-1}. The device's specific capacitance improved by 112.2% (382 F g^{-1}) and its energy density climbed to 13.26 Wh kg^{-1} as a result of the IC's reversible redox reaction. Moreover, it has outstanding cycling stability, with 80.3% of the capacitance remaining after 3000 cycles [116].

The combination of alizarin red S (ARS) with polyvinyl alcohol-sulfuric acid (PVA-H_2SO_4) produced a novel electrolyte known as PVA-H_2SO_4-ARS. Its conductivity reached 33.3 mS cm^{-1}, since ARS in the electrolyte functions as a redox shuttle. The specific capacitance of SCs using a PVA-H_2SO_4-ARS gel-polymer electrolyte is greater (441 F g^{-1} at 0.5 A g^{-1}) than that of ARS-free SCs (160 F g^{-1} at 0.5 A g^{-1}). In addition, it exhibits good cycle stability and an energy density of up to 39.4 Wh kg^{-1}. As a result, there is a good chance that the redox-mediated electrolyte will be used to enhance the electrochemical performance of SCs [119, 123]. By incorporating PySH into PVA-H_3PO_4, Sun *et al* (2015(a)) created a redox-mediated gel-polymer known as polyvinyl alcohol-orthophosphate 2-mercaptopyridine (PVA-H3PO4-PySH) [124]. The PVA-H_3PO_4-PySH system's ionic conductivity rose by 92% to 22.57 mS cm^{-1}. Consequently, a high energy density (39.17 Wh kg^{-1}) and specific capacitance (1128 F g^{-1}) were achieved. According to Ye *et al* (2018), the redox interaction between PySH and the 2,2′-bipyridine redox pair in PVA-H_3PO_4-PySH is responsible for these enhanced characteristics. These findings definitely suggest that redox-mediated gel polymers are interesting options for enhanced flexible solar cells in terms of electrolyte possibilities [125, 126].

Redox electrolytes have significantly improved the performance of SCs, but it is important to remember that most of them have a fatal vulnerability called self-discharge (SD). This issue has been the subject of numerous studies recently [123]

By adding Li_2SO_4-BMIMBr-carbon nanotubes to the PVA solution, Fan *et al* reduced self-discharge and enhanced energy density and cycling stability (capacitance retention 87.9% after 10 000 cycles). The 3D carbon nanotube networks in this solution provide quick ion transmission routes [127]. Chen *et al* used a $CuSO_4$ active electrolyte to suppress the shuffle effect or a Nafion R 177 membrane to inhibit the BQ shuttle in order to block the migration of the active electrolyte between two electrodes and prevent self-discharge [95]. These findings are anticipated to direct future research toward the development of SCs with high energy density and superior energy retention. Ultimately, table 4.1 provides a summary and comparison of a number of common redox electrolyte-based SCs.

4.5 Catholyte and anolyte-based redox electrolyte

The catholyte and anolyte-based energy storage systems have shown the pioneering research fields which have tremendous potential to combine the energy and power density together and mitigate the self-discharge, because in this system the selected ions can transfer through the membrane. The types of membrane have been described in previous sections. The catholytes and analytes work in separate electrodes, i.e. catholyte in positive electrode and anolyte in negative electrode. During charging, the catholyte is oxidized, whereas the anolyte is reduced and during discharge the reverse phenomenon occurs. The electrons flow from the electrode–electrolyte interface to the circuit via the current collector. The point to be noted is that when a single redox species (either catholyte or anolyte) is used to enhance the charge storage system, the performance improvement can be minimal or may be modest as compare to EDLC. However, redox species and also electrodes combined have an outstanding capability to store charge via redox reactions. In contrast, the other counterelectrode can store charge via double-layer formation, resulting in a charge potential imbalance that can be generated. To balance the charge capacity, the electrode mass should be increased. Then the electrode size significantly mismatches and mass-normalized specific energy density will decrease. Additionally, the potential of the counter electrode will achieve the margin of the electrolyte dissociation voltage before the redox-active electrode is fully charged. Consequently, in order to improve the device performance of two separate redox couples is very effective and to prevent the miscibility an ion-exchange membrane or selected atoms exchange membrane is highly required. These two different redox couple arrangements are also known as dual-redox energy storage devices.

The most significant parameters for designing dual-redox ECs are the proper selection of redox-active electrolytes with respect to electrode, pH, and selection of separator membrane with respect to redox species [9, 13]. In the case of redox species, it should have high solubility, reversible electron-transfer capability in the range of operating potential window, chemical and electrochemical stability during intermixed, chemical competence with active components (electrode materials) and passive cell components (current collectors and separators), and excellent cycling stability [16, 24]. Foremost, the standard redox potential, which is close to oxygen- and hydrogen-evolution-potential for the catholyte and the anolyte redox couple respectively, is most desirable. In the above section, we have discussed the redox electrolytes in the light of the redox species nature. But for the fabrication of the cell, the equivalent stable mass charge adjustment is highly desirable for the increase of charge storage capacity. The first dual-redox electrolyte was reported by Frackowiak *et al* whereas $VOSO_4$ was used as anolyte and KI was used as catholyte [42]. Similarly, few dual reported redox couples are viologen/bromide, tin/vanadium, and HQ/methylene blue, etc, which was used in an aqueous system [42, 81, 82, 98]. This system is similar to redox flow batteries, whereas IEMs are effective to block the migration of the redox-active electrolyte in liquid phase. But in this approach, the non-flow redox system and its mechanism are considered for further discussion. However, the basic principle of the redox flow batteries we have briefly described above. Because redox flow batteries are operating in different architectures other passive systems are required, such as storage tanks, pumps, flow meter etc. Herein, we have discussed the chemical materials for the development of redox active electrolytes for the enhancement of hybrid supercapacitor systems.

The redox couples are also used in non-aqueous systems and solid/gel-polymer electrolyte mediums. For example, chloromanganates ($[Mn^{III}Cl_5]^{2-}$/$[Mn^{IV}Cl_6]^{2-}$) salts were added with 1-butyl-1-methylpyrrolidinium ($[BMP]^+$) and 1-ethyl-1-methylpyrrolidinium ($[EMP]^+$) cations. In this system reversible oxidation is happening at the positive electrode by conversion of $[Mn^{II}Cl_4]^{2-}$ to $[Mn^{III}Cl_5]^{2-}$. In contrast, the manganese metal is deposited from $[Mn^{II}Cl_4]^{2-}$ at the negative electrode and produces the resulting cell voltage 2.59 V.

Interestingly, Yu *et al* introduced a new 'mediator ion' concept for the advancement of solid state electrolyte [37]. In this system, they brilliantly introduced the 'solid–electrolyte separator', which is rich in particular ions and placed between the catholyte and anolyte. The exceptionality of the 'mediator ion' strategy is the shuttling of the mediator ion during the redox reactions at the anode and the cathode, due to the presence of the anolyte and the catholyte, respectively. Furthermore, to improve the ion conductivity and also increase the redox species solubility different liquid electrolytes (including different pH) can be employed, such as Zn(KOH/LiOH)|| Li-SSE || $KMnO_4(H_2SO_4)$ is functioning with an acidic-alkaline binary electrolyte. The Li-SSE is $Li_{1+x+y}Al_xTi_{2-x}Si_yP_{3-y}O_{12}$ (LATP) ceramic solid electrolyte membrane, which separates acidic catholyte and alkaline anolyte. During the redox reactions Li^+-ions migrated through the LATP membrane and successfully provide a high-voltage ~ 2.8 V [12, 37, 45]. Several similar reports have been published successfully on the basis of 'mediator-ions' strategy such as $Zn(Zn(NO_3)_2)$ || Li-SSE || $Cu(LiNO_3)$ battery, Zn(LiOH) || Li-SSE || air (H_3PO_4/LiH_2PO_4) and Zn(LiOH) || Li-SSE || Br_2(LiBr).

Not withstanding the many reports that have been published over the last decades which have reported the advantages, dual-redox based hybrid capacitors have not been commercialized due to several hurdles. Firstly, the fabrication of two different electrolytes in a single cell is an industrial challenge, however, for the redox flow batteries two different electrolyte are used, which is used for grid level energy storage applications. But redox flow battery systems are organized with other passive equipment. But in non-flow system it is a major hurdle.

4.6 Conclusion

In the above review, it was observed that appropriate redox electrolytes with suitable porous electrodes are effective combinations for enhancing the charge density and energy storage capacities of hybrid ECs. Particularly, the internal and external surfaces of the porous electrode are extremely operative for the charge storage performances via various forms of redox species adsorption and its entrapment. Additionally, various other factors such as electrode material's properties (electrical conductivity, polarity, chemical stability), applied potential, temperature, electrolytes physicochemical nature, and pH have been playing the dynamic role for the improvement of performance. The structural shortcomings can be mitigated by selecting the proper redox electrolytes species. Additionally, to get effective performances, porous carbon structures are very effective, because they are playing crucial roles regarding redox kinetics, ion diffusion, and adsorption. The porous carbon helps to provide enough space for fast redox reactions by restricting diffusionless redox kinetics.

Furthermore, to increase charge capacity the redox electrolyte operates in a battery-like working principle which can be observed via the non-linear shape of cyclic voltammograms and potentiometric charge–discharge graphs. The point to be noted is that the 'charge to voltage ratio' or 'the charge to potential ratio of an electrode in a redox electrolyte' of a HC with redox electrolytes significantly depends on the applied voltage. Therefore, selection of electrolytes and proper redox species are highly desirable for improved performance with proper safety.

References

[1] Crowley B 2000 Statistical review of world energy *Hydrocarb. Process.* **79** 23
[2] Michelson E S 2016 Which way for the world? A review of energy, economic growth, and geopolitical futures: eight long-range scenarios *Futures* **78–79** 71–3
[3] Lianos P 2017 Review of recent trends in photoelectrocatalytic conversion of solar energy to electricity and hydrogen *Appl. Catal.* B **210** 235–54
[4] Al-Falahi M D A, Jayasinghe S D G and Enshaei H 2017 A review on recent size optimization methodologies for standalone solar and wind hybrid renewable energy system *Energ. Convers. Manage.* **143** 252–74
[5] Kannan N and Vakeesan D 2016 Solar energy for future world: a review *Renew. Sust. Energ. Rev.* **62** 1092–105
[6] Linden D and Reddy T B 2001 *Handbook of Batteries* (McGraw-Hill Education)
[7] Lukatskaya M R, Dunn B and Gogotsi Y 2016 Multidimensional materials and device architectures for future hybrid energy storage *Nat. Commun.* **7** 12647
[8] Chen G Z 2017 Supercapacitor and supercapattery as emerging electrochemical energy stores *Int. Mater. Rev.* **62** 173–202
[9] Evanko B, Boettcher S W, Yoo S J and Stucky G D 2017 Redox-enhanced electrochemical capacitors: status, opportunity, and best practices for performance evaluation *ACS Energy Lett.* **2** 2581–90
[10] Leung P, Martin T, Liras M, Berenguer A M, Marcilla R *et al* 2017 Cyclohexanedione as the negative electrode reaction for aqueous organic redox flow batteries *Appl. Energy* **197** 318–26
[11] Reddy A L M, Srivastava A, Gowda S R, Gullapalli H, Dubey M and Ajayan P M 2010 Synthesis of nitrogen-doped graphene films for lithium battery application *ACS Nano* **4** 6337–42
[12] Aravindan V, Gnanaraj J, Lee Y S and Madhavi S 2014 Insertion-type electrodes for nonaqueous Li-ion capacitors *Chem. Rev.* **114** 11619–35
[13] Senthilkumar S T, Selvan R K and Melo J S 2013 Redox additive/active electrolytes: a novel approach to enhance the performance of supercapacitors *J. Mater. Chem.* A **1** 12386–94
[14] Mai L, Yan M and Zhao Y 2017 Track batteries degrading in real time *Nature* **546** 469–70
[15] Chen Y, Kang Y, Zhao Y, Wang L, Liu J *et al* 2021 A review of lithium-ion battery safety concerns: the issues, strategies, and testing standards *J. Energy Chem.* **59** 83–99
[16] Balasubramaniam S, Mohanty A, Balasingam S K, Kim S J and Ramadoss A 2020 Comprehensive insight into the mechanism, material selection and performance evaluation of supercapatteries *Nano-Micro Lett.* **12:85** 1–46
[17] Salanne M, Rotenberg B, Naoi K, Kaneko K, Taberna P-L, Grey C P, Dunn B and Simon P 2016 Efficient storage mechanisms for building better supercapacitors *Nat. Energy* **1** 16070
[18] Wang Y, Song Y and Xia Y 2016 Electrochemical capacitors: mechanism, materials, systems, characterization and applications *Chem. Soc. Rev.* **45** 5925–50
[19] Béguin F, Presser V, Balducci A and Frackowiak E 2014 Carbons and electrolytes for advanced supercapacitors *Adv. Mater.* **26** 2219–51
[20] Conway B E and Gileadi E 1962 Kinetic theory of pseudo-capacitance and electrode reactions at appreciable surface coverage *Trans. Faraday Soc.* **58** 2493–509
[21] Conway B E 1999 *Electrochemical Capacitors: Scientific Fundamentals and Technological Applications* (New York: Kluwer Academic/Plenum)

[22] Bard A J and Faulkner L R 2001 *Electrochemical Methods: Fundamentals and Applications* (New York: Wiley)
[23] Weingarth D, Foelske-Schmitz A and Kötz R 2013 Cycle versus voltage hold—which is the better stability test for electrochemical double layer capacitors? *J. Power Sources* **225** 84–8
[24] Lee J, Srimuk P, Fleischmann S, Su X, Hatton T A and Pressera V 2019 Redox-electrolytes for non-flow electrochemical energy storage: a critical review and best practice *Prog. Mater Sci.* **101** 46–89
[25] Bandaru P R, Yamada H, Narayanan R and Hoefer M 2015 Charge transfer and storage in nanostructures *Mater. Sci. Eng.* **96** 1–69
[26] Zhang Q, Wang L, Li and Liu G 2020 A real-time energy management control strategy for battery and supercapacitor hybrid energy storage systems of pure electric vehicles *J. Energy Storage* **31** 101721
[27] Lee J, Krüner B, Tolosa A, Sathyamoorthi S, Kim D, Choudhury S, Seo K and Presser V 2016 Tin/vanadium redox electrolyte for battery-like energy storage capacity combined with supercapacitor-like power handling *Energy Environ. Sci.* **9** 3392
[28] Lee J, Choudhury S, Weingarth D, Kim D and Presser V 2016 High performance hybrid energy storage with potassium ferricyanide redox electrolyte *ACS Appl. Mater. Interfaces* **8** 23676–87
[29] Miller E E, Hua Y and Tezel F H 2018 Materials for energy storage: Review of electrode materials and methods of increasing capacitance for supercapacitors *J. Energy Storage* **20** 30–40
[30] Akinwolemiwa B, Peng C and Chen G Z 2015 Redox electrolytes in supercapacitors *J. Electrochem. Soc.* **162** 5054–9
[31] Wang Q, Luo , Hou R, Zaman S, Qi K, Liu H, Park H S and Xia B Y 2019 Redox tuning in crystalline and electronic structure of bimetal–organic frameworks derived cobalt/nickel boride/sulfide for boosted faradaic capacitance *Adv. Mater.* 1905744
[32] Lee J, Srimuk P, Aristizabal K, Kim C, Choudhury S, Nah Y C *et al* 2017 Pseudocapacitive desalination of brackish water and seawater with vanadium pentoxidedecorated multi-walled carbon nanotubes *ChemSusChem.* **10** 3611–23
[33] Lota G and Frackowiak E 2009 Striking capacitance of carbon/iodide interface *Electrochem. Commun.* **11** 87–90
[34] Pal B, Yang S, Ramesh S, Thangadurai V and Jose R 2019 Electrolyte selection for supercapacitive devices:a critical review *Nanoscale Adv.* **1** 3807–35
[35] Chun S-E, Yoo S J and Boettcher S W 2018 Characterization of electric double-layer capacitor with 0.75 M NaI and 0.5 M $VOSO_4$ electrolyte *J. Electrochem. Sci. Technol.* **9** 20–7
[36] Li Z-Y, Akhtar M S, Kwak D-H and Yang O-B 2017 Improvement in the surface properties of activated carbon via steam pretreatment for high performance supercapacitors *Appl. Surf. Sci.* **404** 88–93
[37] Yu X and Manthiram A 2017 Electrochemical energy storage with mediator-ion solid electrolytes *Joule* **1** 453–62
[38] Li B *et al* 2018 Electrode materials, electrolytes, and challenges in nonaqueous lithium-ion capacitors *Adv. Mater.* 1705670
[39] Hubbard A T and Anson F C 1959 Thin-layer chronopotentiometric determination of reactants adsorbed on platinum electrodes *J. Electroanal. Chem.* **9** 163–4
[40] Bae J H, Han J-H and Chung T D 2012 Electrochemistry at nanoporous interfaces: new opportunity for electrocatalysis *Phys. Chem. Chem. Phys.* **14** 448–63

[41] Meller M, Menzel J, Fic K, Gastol D and Frackowiak E 2014 Electrochemical capacitors as attractive power sources *Solid State Ionics* **265** 61–7

[42] Frackowiak E, Fic K, Meller M and Lota G 2012 Electrochemistry serving people and nature: high-energy ecocapacitors based on redox-active electrolytes *ChemSusChem.* **5** 1181–5

[43] Gamby J, Taberna P, Simon P, Fauvarque J and Chesneau M 2001 Studies and characterisations of various activated carbons used for carbon/carbon supercapacitors *J. Power Sources* **101** 109–16

[44] Su L H, Zhang X G, Mi C H, Gao B and Liu Y 2009 Improvement of the capacitive performances for Co–Al layered double hydroxide by adding hexacyanoferrate into the electrolyte *Phys. Chem. Chem. Phys.* **11** 2195–202

[45] Yu H, Wu J, Fan L, Xu K, Zhong X, Lin Y and Lin J 2011 Improvement of the performance for quasi-solid-state supercapacitor by using PVA–KOH–KI polymer gel electrolyte *Electrochim. Acta* **56** 6881–6

[46] Tsutsumi H and Matsuda Y 1993 Electrochemical behavior of ferrocene-deoxycholic acid 1:2 inclusion compound and its application to positive material of zinc-iodine secondary batteries *Electrochim. Acta* **38** 1373–5

[47] Senthilkumar S, Selvan R K, Lee Y and Melo J 2013 Electric double layer capacitor and its improved specific capacitance using redox additive electrolyte *J. Mater. Chem.* A **1** 1086–95

[48] Zhang Y, Zu L, Lian H, Hu Z, Jiang Y, Liu Y *et al* 2017 An ultrahigh performance supercapacitor based on simultaneous redox in both electrode and electrolyte *J Alloy Compd* **694** 136–44

[49] Sankar K V and Kalai S R 2015 Improved electrochemical performances of reduced graphene oxide based supercapacitor using redox additive electrolyte *Carbon* **90** 260–73

[50] Gorska B, Bujewska P and Fic K 2017 Thiocyanates as attractive redox-active electrolytes for high-energy and environmentally-friendly electrochemical capacitors *Phys. Chem. Chem. Phys.* **19** 7923–35

[51] Cha S M, Nagaraju G, Sekhar S C and Yu J S 2017 A facile drop-casting approach to nanostructured copper oxide-painted conductive woven textile as binder-free electrode for improved energy storage performance in redox-additive electrolyte *J. Mater. Chem.* A **5** 2224–34

[52] Lamiel C, Lee Y R, Cho M H, Tuma D and Shim J-J 2017 Enhanced electrochemical performance of nickel-cobalt-oxide@reduced graphene oxide//activated carbon asymmetric supercapacitors by the addition of a redox-active electrolyte *J. Colloid Interface Sci.* **507** 300–9

[53] Chodankar N R, Dubal D P, Lokhande A C, Patil A M, Kim J H and Lokhande C D 2016 An innovative concept of use of redox-active electrolyte in asymmetric capacitor based on MWCNTs/MnO_2 and Fe_2O_3 thin films *Sci Rep.* **6** 39205

[54] Shanmugavani A, Kaviselvi S, Sankar K V and Selvan R K 2015 Enhanced electrochemical performances of PANI using redox additive of K4 [Fe(CN)6] in aqueous electrolyte for symmetric supercapacitors *Mater. Res. Bull.* **62** 161–7

[55] Roldán S, González Z, Blanco C, Granda M, Menéndez R and Santamaría R 2011 Redox-active electrolyte for carbon nanotube-based electric double layer capacitors *Electrochim. Acta* **56** 3401–5

[56] Pang L and Wang H 2021 Inorganic aqueous anionic redox liquid electrolyte for supercapacitors *Adv. Mater. Technol.* **7** 2100501

[57] Wang F, Wu X, Yuan X, Liu Z, Zhang Y, Fu L, Zhu Y, Zhou Q, Wu Y and Huang W 2017 Latest advances in supercapacitors: from new electrode materials to novel device designs *Chem. Soc. Rev.* **46** 6816–54
[58] Evanko B, Yoo S J, Lipton J, Chun S-E, Moskovits M, Ji X, Boettcher S W and Stucky G D 2018 Stackable bipolar pouch cells with corrosion-resistant current collectors enable high-power aqueous electrochemical energy storage *Energy Environ. Sci.* **11** 2865–75
[59] Wu M C, Jiang H R, Zhang R H, Wei L, Chan K Y and Zhao T S 2019 N-doped graphene nanoplatelets as a highly active catalyst for Br_2/Br^- redox reactions in zinc-bromine flow batteries *Electrochim. Acta* **318** 69–75
[60] Li Q, Haque M, Kuzmenko V, Ramani N, Lundgren P, Smith A D, Enoksson P and Power Sources J 2017 Redox enhanced energy storage in an aqueous high-voltage electrochemical capacitor with a potassium bromide electrolyte *J. Power Sources* **348** 219–28
[61] Tang X, Lui Y H, Chen B and Hu S 2017 Functionalized carbon nanotube based hybrid electrochemical capacitors using neutral bromide redox-active electrolyte for enhancing energy density *J. Power Sources* **352** 118–26
[62] Yu F, Zhang C, Wang F, Gu Y, Zhang P, Waclawik E R, Du A, Ostrikov K and Wang H 2020 A zinc bromine 'supercapattery' system combining triple functions of capacitive, pseudocapacitive and battery-type charge storage *Mater. Horiz.* **7** 495–503
[63] Wang Y, Chang Z, Qian M, Zhang Z, Lin J and Huang F 2019 Enhanced specific capacitance by a new dual redox-active electrolyte in activated carbon-based supercapacitors *Carbon* **143** 300–8
[64] Xu Z, Fan Q, Li Y, Wang J and Lund P D 2020 Review of zinc dendrite formation in zinc bromine redox flow battery *Renew. Sustain. Energy Rev.* **127** 109838
[65] Wu M C, Zhao T S, Jiang H R, Zeng Y K and Ren Y X 2017 High-performance zinc bromine flow battery via improved design of electrolyte and electrode *J. Power Sources* **355** 62–8
[66] Yoo S J, Evanko B, Wang X, Romelczyk M, Taylor A, Ji X, Boettcher S W and Stucky G D 2017 Fundamentally addressing bromine storage through reversible solid-state confinement in porous carbon electrodes: design of a high-performance dual-redox electrochemical capacitor *J. Am. Chem. Soc.* **139** 9985–93
[67] Chen S and Zhang J 2020 Redox reactions of halogen for reversible electrochemical energy storage *Dalton Trans.* **49** 9929–34
[68] Bin L, Zimin N, Vijayakumar M, Guosheng L, Jun L, Vincent S and Wei W 2015 Ambipolar zinc-polyiodide electrolyte for a high-energy density aqueous redox flow battery *Nat. Commun.* **6** 6303
[69] Frackowiak E, Meller M, Menzel J, Gastol D and Fic K 2014 *Faraday Discuss.* **172** 179
[70] Lota G, Fic K and Frackowiak E 2011 Alkali metal iodide/carbon interface as a source of pseudocapacitance *Electrochem. Commun.* **13** 38–41
[71] Svensson P H, Kloo L and Synthesis 2003 Structure, and bonding in polyiodide and metal iodide–iodine systems *Chem. Rev.* 103**5** 1649–84
[72] Jiang Y, Cui X, Zu L, Hu Z, Gan J, Lian H, Liu Y and Xing G 2015 High rate performance nanocomposite electrode of mesoporous manganese dioxide/silver nanowires in KI electrolytes *Nanomaterials* **5** 1638
[73] Senthilkumar S T, Selvan R K, Ulaganathan M and Melo J S 2014 Fabrication of Bi2 O 3 ||AC asymmetric supercapacitor with redox additive aqueous electrolyte and its improved electrochemical performances *Electrochim. Acta* **115** 518–24

[74] Qian L, Tian X, Yang L, Mao J, Yuan H and Xiao D 2013 High specific capacitance of CuS nanotubes in redox active polysulfide electrolyte *RSC Adv.* **3** 1703–8
[75] Manan N S A, Aldous L, Alias Y, Murray P, Yellowlees L J, Lagunas M C and Hardacre C 2011 Electrochemistry of sulfur and polysulfides in ionic liquids *J. Phys. Chem.* B **115** 13873–9
[76] Kiyonaga T, Akita T and Tada H 2009 Au nanoparticle electrocatalysis in a photoelectrochemical solar cell using CdS quantum dot-sensitized TiO_2 photoelectrodes *Chem. Commun.* 2011
[77] Yang Z, Chen C Y, Liu C W and Chang H T 2010 Electrocatalytic sulfurelectrodes for CdS/CdSequantum dot-sensitized solar cells *Chem. Commun.* **46** 5485–7
[78] Han W, Kong L B, Liu M C, Wang D, Li J J and Kang L 2015 A high performance redox-mediated electrolyte for improving properties of metal oxides based pseudocapacitive materials *Electrochim. Acta* **186** 478–85
[79] Nagaraju G, Cha S M, Sekhar S C and Yu J S 2016 Metallic layered polyester fabric enabled nickel selenide nanostructures as highly conductive and binderless electrode with superior energy storage performance *Adv. Energy Mater.* 1601362
[80] Díaz P, González Z, Santamaría R, Granda M, Menéndez R and Blanco C 2015 Enhanced energy density of carbon-based supercapacitors using Cerium (III) sulphate as inorganic redox electrolyte *Electrochim. Acta* **168** 277–84
[81] Roldán S, Granda M, Menéndez R, Santamaría R and Blanco C 2012 Supercapacitor modified with methylene blue as redox active electrolyte *Electrochim. Acta* **83** 241–6
[82] Sathyamoorthi S, Kanagaraj M, Kathiresan M, Suryanarayanan V and Velayutham D 2016 Ethyl viologen dibromide as a novel dual redox shuttle for supercapacitors *J. Mater. Chem.* A **4** 4562–9
[83] Li Q, Li K, Sun C and Li Y 2007 An investigation of Cu2+ and Fe2+ ions as active materials for electrochemical redox supercapacitors *J. Electroanal. Chem.* **611** 43–50
[84] Mai L-Q, Minhas-Khan A, Tian X, Hercule K M, Zhao Y-L, Lin and Xu X 2013 Synergistic interaction between redox-active electrolyte and binder-free functionalized carbon for ultrahigh supercapacitor performance *Nat. Commun.* **4** 2923
[85] Lee J, Srimuk P, Fleischmann S, Ridder A, Zeiger M and Presser V 2017 Nanoconfinement of redox reactions enables rapid zinc iodide energy storage with high efficiency *J. Mater. Chem.* A **5** 12520–7
[86] Lee J, Tolosa A, Krüner B, Jäckel N, Fleischmann S, Zeiger M *et al* 2017 Asymmetric tin–vanadium redox electrolyte for hybrid energy storage with nanoporous carbon electrodes *Sustain. Energy Fuels* **1** 299–307
[87] Fic K, Meller M and Frackowiak E 2015 Interfacial redox phenomena for enhanced aqueous supercapacitors *J. Electrochem. Soc.* **162** 5140–7
[88] Mousavi M F, Hashemi M, Rahmanifar M S and Noori A 2017 Synergistic effect between redox additive electrolyte and PANI-rGO nanocomposite electrode for high energy and high power supercapacitor *Electrochim. Acta* **28** 290–8
[89] Gastol D, Walkowiak J, Fic K and Frackowiak E 2016 Enhancement of the carbon electrode capacitance by brominated hydroquinones *J. Power Sources* **326** 587–94
[90] Zhu Y, Liu E, Luo Z, Hu T, Liu T, Li Z *et al* 2014 A hydroquinone redox electrolyte for polyaniline/SnO2 supercapacitors *Electrochim. Acta* **118** 106–11
[91] Vonlanthen D, Lazarev P, See K A, Wudl F and Heeger A J 2014 A stable polyaniline-benzoquinone-hydroquinone supercapacitor *Adv. Mater.* **26** 5095–100
[92] Nie Y F, Wang Q, Chen X Y and Zhang Z J 2016 Nitrogen and oxygen functionalized hollow carbon materials: the capacitive enhancement by simply incorporating novel redox additives into H2SO4 electrolyte *J. Power Sources* **320** 140–52

[93] Nie Y F, Wang Q, Chen X Y and Zhang Z J 2016 Synergistic effect of novel redox additives of p-nitroaniline and dimethylglyoxime for highly improving the supercapacitor performances *Phys. Chem. Chem. Phys.* **18** 2718–29
[94] Chen L, Chen Y, Wu J, Wang J, Bai H and Li L 2014 Electrochemical supercapacitor with polymeric active electrolyte *J. Mater. Chem.* A **2** 10526–31
[95] Chen L, Bai H, Huang Z and Li L 2014 Mechanism investigation and suppression of self-discharge in active electrolyte enhanced supercapacitors *Energy Environ. Sci.* **7** 1750–9
[96] Armand M, Endres F, MacFarlane D R, Ohno H and Scrosati B 2009 Ionic-liquid materials for the electrochemical challenges of the future *Nat. Mater.* **8** 621–9
[97] Sun G H, Li K X and Sun C J 2010 Electrochemical performance of electrochemical capacitors using Cu (II)-containing ionic liquid as the electrolyte *Microporous Mesoporous Mater.* **128** 56–61
[98] Yamazaki S, Ito T, Yamagata M and Ishikawa M 2012 Non-aqueous electrochemical capacitor utilizing electrolytic redox reactions of bromide species in ionic liquid *Electrochim. Acta* **86** 294–7
[99] You D, Yin Z X, Ahn Y, Lee S, Yoo J and Kim Y S 2017 Redox-active ionic liquid electrolyte with multi energy storage mechanism for high energy density supercapacitor *RSC Adv.* **7** 55702
[100] Matsumoto K, Hwang J, Kaushik S, Chen C Y and Hagiwara R 2019 Advances in sodium secondary batteries utilizing ionic liquid electrolytes *Energy Environ. Sci.* **12** 3247
[101] Sathyamoorthi S, Suryanarayanan V and Velayutham D J 2015 Organo-redox shuttle promoted protic ionic liquid electrolyte for supercapacitor *Power Sources* **274** 1135
[102] Yoo K, Dive A M, Kazemiabnavi S, Banerjee S and Dutta P 2016 Effects of operating temperature on the electrical performance of a Li-air battery operated with ionic liquid electrolyte *Electrochim. Acta.* **194** 317
[103] Watanabe M, Thomas M L, Zhang S G, Ueno K, Yasuda T and Dokko K 2017 Application of ionic liquids to energy storage and conversion materials and devices *Chem. Rev.* **117** 7190
[104] Zhang L, Yang S H, Chang J, Zhao D G, Wang J Q, Yang C and Cao B Q 2020 A Review of redox electrolytes for supercapacitors *Front. Chem.* **8** 413
[105] Eftekhari A 2017 Supercapacitors utilising ionic liquids *Energy Storage Mater.* **9** 47
[106] Pan S S, Yao M, Zhang J H, Li B S, Xing C X, Song X L, Su P P and Zhang H T 2020 Recognition of ionic liquids as high-voltage electrolyte for supercapacitors *Front. Chem.* **8** 261
[107] Shahzad S, Shah A, Kowsari E, Iftikhar F J, Nawab A, Piro B, Akhter M S, Rana U A and Zou Y J 2019 Ionic liquids as environmentally benign electrolytes for high-performance supercapacitors *Global Chall.* **3** 1800023
[108] Navalpotro P, Palma J, Anderson M and Marcilla R 2016 High performance hybrid supercapacitors by using para-benzoquinone ionic liquid redox electrolyte *J. Power Sources* **306** 711
[109] Sun L, Zhuo K, Chen Y, Du Q, Zhang S and Wang J 2022 Ionic liquid-based redox active electrolytes for supercapacitors *Adv. Funct. Mater.* **32** 2203611
[110] Zhi-yu X, Pu H, Yang L, Shou-peng N and Peng-fei H 2019 Research progress of polymer electrolytes in supercapacitors *J. Mater. Eng.* **47** 71–83
[111] Batisse N and Raymundo-Piñero E 2017 A self-standing hydrogel neutral electrolyte for high voltage and safe flexible supercapacitors *J. Power Sources* **348** 168–74
[112] Qin H and Panzer M J 2017 Chemically cross-linked poly (2-hydroxyethyl methacrylate)-supported deep eutectic solvent gel electrolytes for eco-friendly supercapacitors *ChemElectroChem.* **4** 2556–62

[113] Hui C-y, Kan C-w, Mak C-I and Chau K-h 2019 Flexible energy storage system-an introductory review of textile-based flexible supercapacitors *Processes* **7** 922

[114] Li H, Lv T, Sun H, Qian G, Li N, Yao Y *et al* 2019 Ultrastretchable and superior healable supercapacitors based on a double cross-linked hydrogel electrolyte *Nat. Commun.* **10** 1–8

[115] Alipoori S, Mazinani S, Aboutalebi S H and Sharif F 2020 Review of PVA-based gel polymer electrolytes in flexible solid-state supercapacitors: opportunities and challenges *J. Energy Storage* **27** 101072

[116] Ma G, Dong M, Sun K, Feng E, Peng H and Lei Z 2015 A redox mediator doped gel polymer as an electrolyte and separator for a high performance solid state supercapacitor *J. Mater. Chem.* A **3** 4035–41

[117] Pan S, Deng J, Guan G, Zhang Y, Chen P, Ren J *et al* 2015 A redoxactive gel electrolyte for fiber-shaped supercapacitor with high area specific capacitance *J. Mater. Chem.* A **3** 6286–90

[118] Tu Q-M, Fan L-Q, Pan F, Huang J-L, Gu Y, Lin J-M *et al* 2018 Design of a novel redox-active gel polymer electrolyte with a dual-role ionic liquid for flexible supercapacitors *Electrochim. Acta* **268** 562–8

[119] Sun K, Ran F, Zhao G, Zhu Y, Zheng Y, Ma M *et al* 2016 High energy density of quasi-solid-state supercapacitor based on redox-mediated gel polymer electrolyte *RSC Adv.* **6** 55225–32

[120] Wang Y, Chang Z, Qian M, Zhang Z, Lin J and Huang F 2019 Enhanced specific capacitance by a new dual redox-active electrolyte in activated carbonbased supercapacitors *Carbon* **143** 300–8

[121] Hashemi M, Rahmanifar M S, El-Kady M F, Noori A, Mousavi M F and Kaner R B 2018 The use of an electrocatalytic redox electrolyte for pushing the energy density boundary of a flexible polyaniline electrode to a new limit *Nano Energy* **44** 489–98

[122] Feng E, Ma G, Sun K, Yang Q, Peng H and Lei Z 2016 Toughened redox-active hydrogel as flexible electrolyte and separator applying supercapacitors with superior performance *RSC Adv.* **6** 75896–904

[123] Seok Jang H, Justin Raj C, Lee W-G, Chul Kim B and Hyun Yu K 2016 Enhanced supercapacitive performances of functionalized activated carbon in novel gel polymer electrolytes with ionic liquid redox-mediated poly(vinyl alcohol)/phosphoric acid *RSC Adv.* **6** 75376–83

[124] Sun K, Dong M, Feng E, Peng H, Ma G, Zhao G *et al* 2015a High performance solid state supercapacitor based on a 2-mercaptopyridine redoxmediated gel polymer *RSC Adv.* **5** 22419–25

[125] Ye T, Li D, Liu H, She X, Xia Y, Zhang S *et al* 2018 Seaweed biomass-derived flame-retardant gel electrolyte membrane for safe solid-state supercapacitors *Macromolecules* **51** 9360–7

[126] Aljafari B, Alamro T, Ram M K and Takshi A 2019 Polyvinyl alcohol-acid redox active gel electrolytes for electrical double-layer capacitor devices *J. Solid State Electrochem.* **23** 125–33

[127] Fan L-Q, Tu Q-M, Geng C-L, Huang J-L, Gu Y, Lin J-M *et al* 2020 High energy density and low self-discharge of a quasisolid-state supercapacitor with carbon nanotubes incorporated redox-active ionic liquid-based gel polymer electrolyte *Electrochim. Acta* **331** 135425

IOP Publishing

Nanomaterials for Advanced Energy and Power Storage Devices

K R V Subramanian, N Sriraam, T Nageswara Rao and Aravinda C L Rao

Chapter 5

Conclusion on current trends in energy storage systems

Venkatesh Sadhana and Janardan Sannapaneni

Notably, energy storage systems (ESS) associated with renewable energy are playing key roles today. The key advantages of renewable energy are low carbon emission and its naturally greater abundance. The amount of power generated is dependent on natural resources like wind and water. Unfortunately, the cost of these power generation projects is high and not suitable to generate the the current electricity demands of the world [1].

Among various renewable energy systems wind and solar energy systems are producers of low carbon electricity. Nuclear technology sources are often deployed with a pumped hydro storage [2]. The new technologies like carbon capture cannot solve the challenges associated with fossil-based power plants and significant research progress required to overcome the issue. The literature has shown that carbon emission into the atmosphere is drastically increasing yearly by fossil-based power plants [3]. Already, countries like Russia and China are facing the same problems with fossil-based power plants [4, 5]. Hence, extensive remedies must be undertaken in a short time to decarbonize it. It is evident that global energy consumption will continue to increase, and a total energy demand increase up to 25% is predicted for 2040, based on 2018 data [6].

It is time to construct large-scale renewable energy storage systems in which electricity is the central form of energy to overcome the environmental crisis caused by the deployment of fossil fuels. Its applications could be broadly covered in power systems such as generation, transmission, distribution, and utilization. These applications have advantages such as, improving the power grid's efficacy, lowering construction cost of generation and power systems, refining power quality and energy efficiency, ensuring a high-quality, safe, realistic power supply, etc [7–11]. The development and commercialization of ESS will have a substantial effect on future power system models [12]. Current data reveals that both engineering and

doi:10.1088/978-0-7503-4901-7ch5

academic research institutes have collaborated and grown at a rapid pace which facilitates the conversion of ESS from small scale industry to large-scale.

A recent International Energy Agency (IEA) published the report 'World Energy Outlook 2016' (WEO 2016) which proved that the pledges made at the 2015 Paris agreement by 195 nations (except US nation) had succeeded in decreasing the CO_2 emissions but will fail to restrict the total emissions to 450 ppmv CO_2 equivalents (so-called the 450 *scenario*) [13]. To control the global temperature, rise to within 2 °C above pre-industrial levels by 2100, the WEO report claims that we need to reach carbon-free energy by the end of this century, whereas renewable energy sources (RES) need to supply more than 60% of global energy (other supplementary actions like 715 million electrical vehicles need to be deployed on the road by 2040). To achieve these major targets, we will have to enhance in areas like science and technology, which in turn affects the financial status of every nation [14]. Conventional energy production and distribution systems are unable to maintain the balance between energy demand and supply on the electrical grid system. Moreover, the generated energy is often wasted due to lack of proper storage capacity and mechanism.

The basic principle of any ESS comprises transforming one type of energy into another type of energy that can, whenever required, render back the stored energy with high efficiency, in a cost effective and reliable manner. Nevertheless, rapid and large-scale deployment of intermittent sources like wind, tide and wave, and solar energy depends on the amount of energy produced, cost effectiveness and efficiency. As Bloomberg Technology News stated on July 31, 2017, Germany had to discard 4% of wind energy in 2015 and China has abandoned 17% of its renewable energy, while California had to give up 300 000 MWh of wind and solar energy in the first half of 2017, all due to shortage of adequate electricity storage capacity [15].

A broad classification in ESS (current trends) has been developed based on the way energy is stored. Among them the foremost ESS is mechanical, followed by heat energy, electrochemical, magnetic, and lastly chemical energy storage [16].

Mechanical energy storage (MES): MES is broadly classified into three systems: pumped storage, compressed-air energy storage, and flywheel energy storage. Among these pumped storage is the oldest technique which has advantages such as substantial capacity, long service lifespan, and minimal unit cost. On the other hand, the pumped storage system is restricted to terrestrial conditions. Apart from that, time to construct and cost is greater. The compressed storage system has the advantages of very long operation time, long lifespan and moreover, is capable of delivering combined heating, cooling, and electricity by the process of energy transformation (compressed air into other alternative energy sources). However, energy production is lower and it requires air storage mine tunnels which are more expensive [17, 18]. The energy efficacy problem is overcome by implementing the advanced method flywheel energy storage system. A few other advantages are the low maintenance costs, long service lifespan, good stability, eco-friendly in nature, easy to self-discharge, but the energy density is lower [19, 20]. The stored kinetic energy is converted to electrical energy by reducing the speed of the flywheel by a torque. Applications include the amalgamation of flywheel energy storage systems with a renewable energy storage power plant system [21].

Recent, research found that FES's flywheel energy storage system have good features, with a high efficiency of 90% [22].

Heat energy storage (HES): HES systems are remarkably the most used ESS around the globe, as the US department of energy reports stated that HES accounted for 1.9% (capacity 3.3 GW) of the world's energy storage in 2017 [23]. Heat energy systems essentially support both domestic and commercial energy usage, and thus participate in management of the respective microgrid [24]. HES applications are used by both heat and cold storage media using distinct conditions, namely temperature, place, and power. The applications are broadly classified into: latent heat, absorption, sensible heat, and adsorption system [25, 26].

Depending on the operating temperature, HES systems are divided into two types: low temperature energy storage (LTES) systems and high temperature energy storage (HTES) systems [27, 28]. One of the disadvantages of this systems is the large space requirement. The energy storage efficiency can reach more than 95% and the production cost is about 1/30 of the large-scale battery systems [29].

Electrochemical energy storage (EcES): This is one of the oldest storage systems, it relies on the reversible electrochemical reaction used for producing and storing of DC power. All secondary batteries (rechargeable) and flow batteries (redox) store the electrical energy in the form of chemical energy which comes under EcES [30]. It has wide range of electrical energy storage capacities in the range of 10 Wh kg^{-1} to 13 KW kg^{-1} with an efficacy of 70%–80% with the least maintenance without carbon emission [31, 32]. Due to the various sizes available and easy transportation features, EcES systems have commercially succeeded in both residential and commercial scale utility applications [33].

Depending on fuel and electrode materials, EcES is classified into two types namely, battery energy storage (BES) systems and flow battery energy storage (FBES) systems [34]. BES rely on electrochemical reactions (redox) that convert chemical energy into electrical energy and vice versa. Based on the charging phenomenon, there are two types namely, primary and secondary batteries which are non-rechargeable and rechargeable respectively [35]. A few of the secondary batteries are listed below as lead–acid (LA), lithium–ion, nickel–cadmium (Ni–Cd), sodium sulfur (NaS), lithium–ion (Li–ion), sodium–ion (Na–ion), and metal–air batteries, depending on the electrode's material and electrolyte.

The flexible range of power densities (90 kW m^{-3} to 10 MW m^{-3}) and energy densities (75–800 kWh m^{-3}) exhibited by BES enables most of the stationary and mobile applications to be supported [36]. However, the advantages of BES mentioned in the literature are: long lifetime, high energy source, available in diverse sizes depending on requirements [37]. Many BES suffer from challenges like toxic electrode and electrolyte materials, disposal problem, recycling of batteries, and small depth of discharge (DoD) [38]. Further BES systems are classified into storage integrated BES systems (e.g. lead–acid, lithium–ion, sodium–sulfur, nickel–cadmium, and nickel–metal hydrate batteries) and external storage BES systems (e.g. vanadium redox flow, ZnBr and Zn–air systems).

On the other hand, flow batteries are equipped with advanced aqueous electrolyte systems having the advantages of operating at wide range of temperatures, e.g. near to ambient [39]. The working mechanism of flow batteries deviate from secondary batteries where the electrode materials participate in redox reactions, in flow batteries the energy is stored in active electrolytes (two separate external tanks), hence no self-discharge [40]. Other advantages including low self-discharge, response time in milli seconds, room temperature operations, long charge and discharge lifecycles, capable of high overloading for short times, etc [41].

Electromagnetic energy storage (EmES): These systems directly store energy in an electric field without converting into other energy forms [42]. EmES systems are broadly classified into electrostatic energy storage systems and magnetic energy storage systems. The capacitors and super-capacitors come under the category of electrostatic energy storage systems (SCES—super capacitor energy storage) and superconducting magnetic energy storage (SMES) is a magnetic energy storage system. The advantages of super-capacitors include fast response, high efficiency, long cycle life, high-power density, minimal maintenance, wide operational temperature range, etc. Nonetheless, because of low energy density, super capacitors are suitable for application in combination with other energy storage technologies [43, 44]. They are highly suitable for high energy requirement. However, there are several drawbacks such as inflated cost, low energy density and complex maintenance [45]. SCES systems are also referred to as ultra-capacitor energy storage (UCES) or electric double layer capacitors (EDLC). They store energy in the form of an electrostatic field [46] and have additional advantages like high specific energy, low internal resistance, and wide temperature range, which aims to replace conventional capacitors employed in power electronics and storage batteries [47, 48].

SMES stores energy in the form of the magnetic field of a superconducting coil due to flow of direct current [49]. Depending on the superconductor material and cryogenic conditioning system the SMES are categorized into two groups: low temperature superconductors (LTS): SMES systems built with NbTi superconductors and liquid helium coolant at 4.3 K and high temperature superconductors; (HTS) SMES systems are built from ceramic oxide superconductors and liquid nitrogen coolant at 77 K. HTS SMES systems are cost efficient and less expensive than the LTS SMES systems [50]. The advantages of SMES systems are stated as quick response of charge and discharge of very high-power density (up to 5000 MWh) and high efficiency of the order 95%–98% having an approximate lifetime of 30 years [51]. Other advantages are that it can admit the ramp and transients at a very quick rate, thereby the power system units can be functioning at maximum efficiency optimum points.

Likewise, in microgrid applications SMES systems can be used to exchange real and reactive powers. Various hybrid SMES applications in distributed grid architectures and transport machinery are under development for the design of vehicular and current source grid inverters [52, 53].

Chemical energy storage (CES): The CES system represents a green technology which produces and stores large-scale energy without any pollution. The energy storage amount is greater than 100 GWh. These systems are more efficient for the long term and the energy is stored in the form of chemical bonds and the energy is produced from the chemical reaction called molecular rearrangements [54]. The chemical fuel industry has received so much attention because of its production and transportation of electricity worldwide. Examples of fuels are: butane, ethane, diesel, natural gas, coal, gasoline, liquefied petroleum gas (LPG), propane, and hydrogen. Initially, these fuels are transformed to mechanical energy and subsequently into electricity [35]. CES systems essentially focused on hydrogen, synthetic natural gas, and solar fuel storage systems [55]. Nevertheless, CES systems have disadvantages like, low energy conversion efficacy (40%–50%), high cost, high investment, and minimal safety [56]. Currently, hydrogen energy storage systems have received more attention all over the world. The fuel cell (FC) is the basic principle behind the electrical energy generation from chemical energy. The mechanism of the FC depends on the amount of fuel and oxidant substances supplied. It can yield electricity, if externally supplied active fuel and oxidant are available and it even helps in reducing the emission of harmful substances released from fossil fuels [57].

5.1 Hydrogen energy storage system

In all energy storage systems, hydrogen is esteemed with clean energy and carbon-free emission chemical energy systems [35, 54]. Hydrogen gas can be produced from the water electrolysis process (acid/alkaline) or directly from sunlight, using photocatalytic water splitting mechanisms. When water is undergoing the electrolysis process hydrogen gas is liberated and stored in hydrogen tanks and the process is said to be charging. During the discharging process, the energy is generated in the form of electrons using the fuel cell redox mechanism (oxidation at anode and reduction at cathode). An electrolyzer is a physical device which breaks down water into hydrogen and oxygen and the hydrogen is stored in high pressurized tanks and oxygen is liberated into the atmosphere [58]. The fuel cell is basically composed of: anode, cathode, electrolyte, and separator. The anode reaction is oxidation where hydrogen gas is oxidized into protons and electrons transfer through the external circuit. The reactants like protons, oxygen (oxidant), and electrons react at the cathode to produce water and heat. During these redox reactions the electrons flow through an external circuit, finally electricity is produced. The hydrogen fuel cell is broadly classified into six types namely, alkaline fuel cell (AFC), phosphoric acid fuel cell (PAFC), solid polymer fuel cell proton exchange membrane FC (SPFC-PEMFC), solid oxide fuel cell (SOFC), molten carbonate fuel cell (MCFC), and direct methanol fuel cell (DMFC) [59]. A substantial amount of energy can be stored in a FC by optimizing the size of the hydrogen tank. The applications of FC are in various fields like microgrid power generation and automobile industry. The biggest advantage of FCs is the emission of water vapor into the environment. The energy carriers in these systems, hydrogen and methane, can be synthesized by electrolyzed hydrogen reaction with carbon dioxide [60]. Fuel cells are direct and indirect systems depending on the direct reaction of fuel and indirect transition into enriched hydrogen gas before entering the redox reaction chamber [61]. Recent studies have shown that advances like solid-state hydrogen storage in metal hydrides and current novel discoveries are moving toward a hydrogen economy. Recent innovations in electrode materials are reported like Mg–Li–Al and Mg–Na–Al systems employed for solid-state hydrogen storage function [62, 63].

5.1.1 Synthetic natural gas (SNG)

Apart from hydrogen gas, natural gas is the most widely used clean energy source and dominates energy supply. An efficient alternative energy production system is the conversion of coal to synthetic natural gas which has proved to be sustainable [27]. The production of biomass is completely from plants and animal waste with no carbon emission [64, 65]. SNG can be stored in pressure tanks, underground caverns, or fed directly into the gas grid. The end users of SNG systems are residential homes. Electricity can also be generated from biomass, with the help of a generator.

5.1.2 Solar fuels

The aim of solar fuels is to trap the abundant sunlight (solar energy) and convert it into a useful and storable chemical energy resource. It is a kind of energy conversion and storage system and a completely renewable and green energy source. Recent literature has shown that the methods and advances for producing solar fuel production are: natural photosynthesis, artificial photosynthesis, and thermochemical production [66]. Photosynthesis is a natural phenomenon, in which the energy is stored in the form of chemical compounds like carbohydrates. All chlorophyll containing plant species will do the same and the energy system requires water and CO_2 as raw materials [67]. The artificial leaf referred to as bionic leaf in which the energy storage process is the same as in a natural leaf. The thermochemical technique uses sunlight to heat materials to extremely high temperatures, where they react with steam or CO_2 to produce carbon monoxide (CO) or hydrogen (H_2) [68]. Since the 1950s, engineers and scientists have been collaborating to scale up the laboratory prototype of solar fuel to commercial scales. Recent articles have revealed the current trends in CO_2 photoreduction, light harvesting and activation of CO_2 molecules including a solution for enhancing photocatalytic activity [69]. Developed countries like USA, Australia, and South Korea have invested in several renewable energy research centers and India, China, and Japan are in the same streamline [67, 70].

5.1.3 Future aspects of energy storage systems

ESS has reached a milestone in the different possibilities in the renewable energy generation grid integration, distributed generation, microgrid, transmission and distribution, smart grid, and ancillary services. Statistical analysis carried out for the development of energy storage in China 2050, showed that the demand for energy storage will range between 560 and 780 GW and energy will accomplish between 2 and 3 billion kWh without considering grid constraints among regions in China.

Nonetheless, the large-scale application of energy storage technology has some major hindrances both in terms of technological and financial circumstances.

Technological challenges exist such as the capacity of ESS's, long-lifespan, less cost of production and safety for electrochemical energy storage systems. Secondly, physical storage technology with high-efficacy and low-cost is expected. Advanced research must be encouraged in the field of storage simulation and operation optimization in multiple applications. On the other hand, the advancement will promote ESS in the field of industrialization as well as commercialization. Finally, the new generation of ESS are multi-diversified and applicable for all energy systems.

Economic constraints of ESS are associated with integration of energy storage systems with renewable energy requiring more investment and add to overall cost of the energy system. Furthermore, investigation must happen in novel storage materials which directly limit the cost of production.

Lack of policies to assist the progress of the technology is also another contributing factor leading to an increase in the total cost of the technology.

References

[1] Dunn B, Kamath H and Tarascon J-M 2011 Electrical energy storage for the grid: a battery of choices *Science* **334** 928–35

[2] Yang C-J and Jackson R B 2011 Opportunities and barriers to pumpedhydro energy storage in the United States *Renew. Sustain. Energy Rev.* **15** 839–44

[3] Scott V, Gilfillan S, Markusson N, Chalmers H and Haszeldine R S 2013 Last chance for carbon capture and storage *Nat. Clim. Change* **3** 105–11

[4] Liu Z 2016 *China's Carbon Emissions Report* (Heidelberg: Springer)

[5] Olivier J, Janssens-Maenhout G, Muntean M and Peters J 2016 *Trends in GlobalCO2Emissions: 2016 Report* (Netherlands: PBL) p 81

[6] IEA, World Energy Outlook 2019 2019. Paris, https://iea.org/reports/world-energy-outlook-2019

[7] Lu Q Y, Hu W, Min Y *et al* 2015 A multi-pattern coordinated optimization strategy of wind power and energy storage system considering temporal dependence *Autom. Electr. Power Syst.* **39** 6–12

[8] Niu Y, Zhang F, Zhang H *et al* 2016 Optimal control strategy and capacity planning of hybrid energy storage system for improving AGC performance of thermal power units *Autom. Electr. Power Syst.* **40** 38–45

[9] Wang H J and Jiang Q Y 2014 An overview of control and configuration of energy storage system used for wind power fluctuation mitigation *Autom. Electr. Power Syst.* **38** 126–35

[10] Yuan X M, Cheng S J and Wen J Y 2013 Prospects analysis of energy storage application in grid integration of large-scale wind power *Autom. Electr. Power Syst.* **37** 14–8

[11] Liu W J, Sun L, Lin Z Z *et al* 2015 Short-period restoration strategy in isolated electrical islands with intermittent energy sources, energy storage systems and electric vehicles *Autom. Electr. Power Syst.* **39** 49–58

[12] Zhou X X, Lu Z X, Liu Y M *et al* 2014 Development models and key technologies of future grid in China *Proc. CSEE* **34** 4999–5007

[13] International Energy Agency, World Energy Outlook 2016 (Nov. 2016), also see http://iea.org/newsroom/news/2016/november/world-energy-outlook-2016

[14] World Energy Outlook 2015 (London: International Energy Agency)

[15] Bloomberg Technology News 2017 https://bloomberg.com/news/articles/2017-07-31/alphabet-wants-to-fixrenewable-energy-s-storage-problem-with-salt

[16] Yu E K and Chen L J 2011 Characteristics and comparison of largescale electric energy storage technologies *Zhejiang Electr. Power* **12** 4–8

[17] Mei S, Wang J, Tian F *et al* 2015 Design and engineering implementation of non-supplementary fired compressed air energy storage system: TICC-500 *Sci. China Technol. Sci.* **58** 600–11
[18] Xue X D, Mei S W, Lin Q Y *et al* 2016 Energy inter-net oriented non-supplementary fired compressed air energy storage and prospective of application *Power Syst. Technol.* **40** 164–71
[19] Chen Y A, Gan S L, Zhou J H *et al* 2016 Energy storage technology of flywheel *Chin. J. Power Sour.* **40** 1718–21
[20] Zhang X B, Chu J W, Li H L *et al* 2015 Key technologies of flywheel energy storage systems and current development status *Energy Storage Sci. Technol.* **4** 55–60
[21] Hadjipaschalis I, Poullikkas A and Efthimiou V 2009 Overview of current and future energy storage technologies for electric power applications *Renew. Sustain. Energy Rev.* **13** 1513–22
[22] Ribeiro P F, Member S, Johnson B K, Crow M L, Member S, Arsoy A *et al* 2001 Energy storage systems for advanced power applications *Proc. IEEE* **89** 1744–56
[23] 2017 *U.S. Energy and Employment Report* (Washington, DC: U.S. Department of Energy) p 84
[24] World Energy Outlook 2011 *OECD, IEA* (Paris: International Energy Agency (IEA) Publications) p 577
[25] Guney M S and Tepe Y 2017 Classification and assessment of energy storage systems *Renew. Sustain. Energy Rev.* **75** 1187–97
[26] Cabeza L F, Martorell I, Miró L, Fernández A I and Barreneche C 2015 *Introduction to Thermal Energy Storage (TES) Systems* (Woodhead Publishing Limited)
[27] Chen H, Cong T, Yang W *et al* 2009 Progress in electrical energy storage system: a critical review *Prog. Nat. Sci.* **19** 219–312
[28] Akinyele D and Rayudu R 2014 Review of energy storage technologies for sustainable power networks *Sustain. Energy Technol. Assess* **8** 74–91
[29] Vick B D and Moss T A 2013 Adding concentrated solar power plants to wind farms to achieve a good utility electrical load match *Sol. Energy* **92** 298–312
[30] International Electrotechnology Commission 2011 Electrical energy storage White paper 1–78
[31] Divya K C and Østergaard J 2009 Battery energy storage technology for power systems: an overview *Elect. Power Syst. Res.* **79** 511–20
[32] Cho J, Jeong S and Kim Y 2015 Commercial and research battery technologies for electrical energy storage applications *Prog. Energy Combust. Sci.* **48** 84101
[33] Kim J, Suharto Y and Daim T U 2017 Evaluation of electrical energy storage (EES) technologies for renewable energy: a case from the US pacific northwest *J. Energy Storage* **11** 2554
[34] Gür T 2018 Review of electrical energy storage technologies, materials and systems: challenges and prospects for large-scale grid storage *Energy Environ. Sci.* **10** 2696–767
[35] Wagner L 2007 Overview of energy storage methods *Research Report*
[36] Chatzivasileiadi A, Ampatzi E and Knight I 2013 Characteristics of electrical energy storage technologies and their applications in buildings *Renew. Sustain. Energy Rev.* **25** 814–30
[37] Sadoway D 2012 The Missing Link to Renewable Energy, TED2012
[38] Review of Electrical Energy Storage Technologies and Systems and of their Potential for the UK 2004 pp 1–34
[39] Reddy T B 2011 *Linden's Handbook of Batteries* 4th edn (New York: McGraw-Hill)

[40] Leung P, Li X, Ponce De León C, Berlouis L, Low C T J and Walsh F C 2012 Progress in redox flow batteries, remaining challenges and their applications in energy storage *RSC Adv.* **2** 1012510156
[41] Guarnieri M, Mattavelli P, Petrone G and Spagnuolo G 2016 Vanadium redox ow batteries: potentials and challenges of an emerging storage technology *IEEE Ind. Electron. Mag.* **10** 20–31
[42] U.S. Department of Energy 2011 *Energy Storage: Program Planning Document* (Washington, DC: U.S. Department of Energy)
[43] Zhang X Y, Zhang H Z, Lin Z Q *et al* 2016 Recent advances and challenges of stretchable supercapacitors based on carbon materials *Sci. China Mater.* **59** 475–94
[44] Be×guin F, Presser V, Balducci A *et al* 2014 Carbons and electrolytes for advanced supercapacitors *Adv. Mater.* **26** 2219–51
[45] Luo X, Wang J, Dooner M *et al* 2015 Overview of current development in electrical energy storage technologies and the application potential in power system operation *Appl. Energy* **137** 511–36
[46] Sahay K and Dwivedi B 2009 Supercapacitors energy storage system for power quality improvement: an overview *Quality* **5** 1–8
[47] Mishra R and Saxena R 2017 Comprehensive review of control schemes for battery and super-capacitor energy storage system *Proc. 7th Int. Conf. Power Syst.* pp 702–7
[48] Zhu C, Lu R, Tian L and Wang Q 2006 The development of an electric bus with super-capacitors as unique energy storage *Proc. IEEE Veh. Power Propuls. Conf. (VPPC)* pp 1–5
[49] Xue X D, Cheng K W E and Sutanto D 2005 Power system applications of superconducting magnetic energy storage systems *Proc. Ind. Appl. Conf.* pp 1524–9
[50] Aware M V and Sutanto D 2003 Improved controller for power conditioner using high-temperature superconducting magnetic energy storage (HTSSMES) *IEEE Trans. Appl. Supercond.* **13** 38–47
[51] Ali M H, Wu B and Dougal R A 2010 An overview of SMES applications in power and energy systems *IEEE Trans. Sustain. Energy* **1** 38–47
[52] Wang Z, Zou Z and Zheng Y 2013 Design and control of a photovoltaic energy and SMES hybrid system with current-source grid inverter *IEEE Trans. Appl. Supercond.* **23** Art. no. 5701505
[53] Trevisani L, Morandi A, Negrini F, Ribani P L and Fabbri M 2009 Cryogenic fuel-cooled SMES for hybrid vehicle application *IEEE Trans. Appl. Supercond.* **19** 2008–11
[54] Revankar S, Bindra H and Revankar S 2019 Chemical energy storage *Storage and Hybridization of Nuclear Energy* 1st edn (Pittsburgh, PA: Academic) pp 177–227
[55] Kampouris K, Drosou V, Karytsas C *et al* 2020 Energy storage systems review and case study in the residential sector *IOP Conf. Ser.: Earth Environ. Sci.* **410** 012033
[56] Huo X X, Wang J, Jiang L *et al* 2016 Review on key technologies and applications of hydrogen energy storage system *Energy Storage Sci. Technol.* **5** 197–203
[57] Eisler M N 2018 The carbon-eating fuel cell [blueprints for a miracle] *IEEE Spectr.* **55** 22–76
[58] Cannolly D 2010 *PhD Thesis* A review of energy storage technologies: for the integration of fluctuating renewable energy
[59] Nadeem F, Hussain S, Tiwari P *et al* 2018 Comparative review of energy storage systems, their roles, and impacts on future power systems *IEEE Access* **7** 4555–85
[60] Yao L *et al* 2013 Challenges and progresses of energy storage technology and its application in power systems *J. Mod. Power Syst. Clean Energy* **4** 519–28

[61] Moore R M 2000 Indirect-methanol and direct-methanol fuel cell vehicles *Proc. 35th Intersoc. Energy Convers. Eng. Conf. Exhib.* vol 2 pp 1306–16

[62] Ali N and Ismail M 2021 Advanced hydrogen storage of the Mg–Na–Al system: a review *J. Magnes. Alloy* **9** 1111–22

[63] Sazelee N, Ali N, Yahya M *et al* 2022 Recent advances on Mg–Li–Al systems for solid-state hydrogen storage: a review *Front. Energy Res.* **10** 875405

[64] Onumaegbu C, Mooney J, Alaswad A and Olabi A G 2018 Pre-treatment methods for production of biofuel from microalgae biomass *Renew. Sustain. Energy Rev.* **93** 16–26

[65] Onumaegbu C, Alaswad A, Rodriguez C and Olabi A 2018 Optimization of pre-treatment process parameters to generate biodiesel from microalga *Energies* **11** 806

[66] Styring S 2012 Solar fuels: vision and concepts *Ambio* **41** 156–62

[67] Heeger A 2012 *Solar Fuels and Artificial Photosynthesis-Science and Innovation to Change our Future Energy Options* (London: Royal Society of Chemistry) pp 1–23

[68] Vriend H and Purchase R 2013 Solar fuels, and artificial photosynthesis, Biosol. https://lisconsult.nl/files/docs/Solar_fuels_final_version.pdf. Cells

[69] Gong E, Ali S, Hiragond C *et al* 2022 Solar fuels: research and development strategies to accelerate photocatalytic CO_2 conversion into hydrocarbon fuels *Energy Environ. Sci.* **15** 880–937

[70] Purchase R, Vriend H, Groot H *et al* 2015 *Artificial Photosynthesis: for the Conversion of Sunlight to Fuel* (Leiden University)

www.ingramcontent.com/pod-product-compliance
Ingram Content Group UK Ltd.
Pitfield, Milton Keynes, MK11 3LW, UK
UKHW051849100926
8556IPUK00005B/529